The Only Woman in the Room

The
Only
Eileen
Pollack
Woman
in the
Room

WHY SCIENCE IS
STILL A BOYS' CLUB

BEACON PRESS
BOSTON

BEACON PRESS
Boston, Massachusetts
www.beacon.org

Beacon Press books
are published under the auspices of
the Unitarian Universalist Association of Congregations.

18 17 16 15   8 7 6 5 4 3 2

This book is printed on acid-free paper that meets the uncoated paper
ANSI/NISO specifications for permanence as revised in 1992.

An excerpt from this book first appeared in the *New York Times Magazine* under
the title "Can You Spot the Real Outlier?," October 6, 2013, pp. 30–35 and 44–46.
Portions of chapter 1 previously appeared under the title "Pigeons" in *Prairie
Schooner* 86, no. 1 (Spring 2012): 62–71, and in *The Best American Essays 2013*,
ed. Cheryl Strayed (New York: Mariner Books, 2013), 113–22. Some of the material
in chapter 6 previously appeared in a different form in the essay "Syllogisms," in
*Between Mothers and Sons: Women Writers Talk about Having Sons and Raising
Men*, ed. Patricia Stevens (New York: Scribner, 1999), 80–98.

Text design and composition by Kim Arney

Library of Congress Cataloging-in-Publication Data
Pollack, Eileen.
    The only woman in the room : why science is still a boys' club / Eileen Pollack.
        pages   cm
    Includes bibliographical references.
    ISBN 978-0-8070-4657-9 (hardback)—ISBN 978-0-8070-4661-6 (ebook)
1. Women scholars—United States. 2. Women scientists—United States. 3. Sex
discrimination against women—United States. 4. Pollack, Eileen, 1956– I. Title.
    HQ1397b .P65 2015
    305.420973—dc23

                                                                    2015004217

To my sister, Joan, and my brother, Sheldon
And in loving memory of my mother,
Wilma Davidson Pollack

CONTENTS

Bright College Years

In July 2008, I stood beneath an elegant blue-and-white-striped tent in a courtyard at Yale and listened in disbelief as my former classmates reminisced about their drunken bouts at Mory's, the regattas in which they had sailed, the fierce squash matches and avant-garde theatricals in which they had taken part, and the As they had accumulated dozing through Rocks for Jocks. As my classmates rose from their seats and waved their white napkins while harmonizing with a chorus of middle-aged Whiffs in a nostalgic refrain of "Bright College Years," I needed to lean against a nearby ivied wall to keep from wobbling.

My bright college years, in the mid-1970s, had been spent in the basement of Kline Science Library, paging frantically through physics textbooks in the hopes of finding problems similar to the ones I needed to solve for that week's assignment; or in the Gothic laboratories on Science Hill, trying to connect a wire to an oscilloscope without electrocuting myself; or in the newly opened Yale Computing Center, where I fed punch cards through the reader and prayed that when I got my printout several days later, the program wouldn't have crashed because I had punched a period instead of a comma or one of the hairs I frequently shed (no doubt from all that stress) had prevented a card from being processed. Unlike most of my classmates, I had majored in science. In fact, I was one of the first two women to earn a bachelor of science degree in physics at Yale. Two of my female classmates pursued the less rigorous but still demanding bachelor of arts in physics, but as a measure of how isolated we all were, I didn't learn of the existence of one of these two women until that day at our reunion. As she and I stood comparing the reasons neither of us had gone on for our PhDs (my classmate had become an ophthalmologist), we couldn't help but wonder what, if anything, had changed in the intervening decades.

The odds that I would have gotten into Yale in the first place were miniscule to nil. I attended a rural public school whose accelerated track in physics and math I wasn't allowed to enter because, as my principal put it, "girls never go on in science and math." Angry and bored, I began staying home from school and sitting in my parents' attic, reading books about space and time. Senior year, after teaching myself calculus, I insisted that the school administer AP exams in the subjects I had studied (as well as English, which I loved but didn't value because everyone seemed to think that teaching English would be a lovely career for a girl like me). My efforts won me a spot at Yale, but I was woefully unprepared. Most of the young men in my classes had attended fancy prep schools like Andover or St. Paul's, or magnet schools like the Bronx High School of Science, where they studied several terms of advanced calculus and one or two years of intensive physics. They yawned as the professor sped through the material while I grew panicked at how little I understood. As the only female student, I debated whether to raise my hand and expose myself to ridicule as the one dim-witted girl who made all the boys wait while she asked a stupid question. By the time I resolved to ask my question anyway, I had lost track of the lecture and slipped even farther behind.

To make matters worse, I had grown up in Liberty, New York, in the heart of the Jewish Catskills, an outrageously profane, working-class resort area, where my grandparents owned a hotel, and I felt out of place at Yale, where even the other Jewish students seemed so aristocratic I had no idea they *were* Jews. Very few women taught on the faculty (none at all in physics or math), and I had only one female friend, my equally overwhelmed pre-med roommate, Laurel. Although no single obstacle caused me to give up or fail, the constant need to jump so many hurdles wore me down.

Take chemistry lab freshman year. The first class fell on Rosh Hashanah. Guilty about missing services, I was equally afraid to miss my first lab. As a compromise, I attended synagogue in the morning and then, still wearing my dress and pantyhose, showed up for the lab. When I followed the TA's instructions and poured a solution of hydrochloric acid into a glass burette, I left the stopcock open and dribbled acid on my legs. My stockings began to smoke; the nylon dissolved in crimson rivulets that ran, steaming, down my calves. Certain that the rivulets were blood but reluctant to call attention to myself by screaming, I hurried down the aisle between my classmates' benches and tapped the lab assistant on his back. "Excuse me," I said. "I've had an accident."

He wheeled around. "You what?"

I explained what I had done, and, to neutralize the acid, he dumped bicarbonate of soda on my legs. Horrified, I looked down at my pantyhose, which now gaped with ragged holes and were clotted with thick white powder. When I looked up, seventeen pairs of male eyes were focused on my legs. The accident could have happened to anyone. Except that, if it had, they would have been wearing pants.

Or consider my relationship to my advisor, Professor Parker. His first name was Peter, and back then, in the era when Peter Parker was a comic-book hero known only to the geeky few, I didn't understand why this made my classmates laugh, or why my professor wore a spider on his belt buckle, or why the door to his office—which could be reached only via the entrance to Yale's nuclear reactor, buried like a pharaoh's tomb in Science Hill—was decorated with a giant web.

"Spider-Man?" I said when a classmate explained the joke. "Who the hell is that?"

The Only Woman in the Room is my long-postponed attempt to understand how and why I worked so hard to earn a bachelor of science degree in physics, only to become a writer. It is an answer to my son, who cannot comprehend why his strong and seemingly self-possessed mother didn't achieve her dream to become a physicist. It is also my answer to Lawrence Summers, the former president of Harvard, whose ill-considered ruminations over why more women don't end up holding tenured positions in the hard sciences, delivered during a lunchtime talk in January 2005, ignited a nationwide furor that simmers on today. I first met Larry in my teens, when he was a championship debater at MIT and judged my high school team. Then as now, he struck me as a man who admires and enjoys the company of smart, independent women. When the controversy over his remarks erupted—he suggested that innate differences in scientific and mathematical aptitude at the very highest end of the spectrum, along with differences in socialization, discrimination, and most women's reluctance to pursue careers that demand intense devotion, might account for the paucity of tenured female faculty in science and engineering—I got the sense he had asked the question because he cared about the answer. I was upset that he had intimated that the scarcity of female physicists could be attributed to innate differences between the sexes. But as I read the heated responses to his comments, I wondered if anyone could understand why so few women end up as tenured physicists unless he had experienced what I had experienced in attempting to become one. I wasn't sure I understood those reasons myself. I sat down to write my friend an e-mail, but the e-mail grew so long that I realized what I needed to do was write this book.

And to write this book, I would need to revisit the painful memories of my own childhood love for science, my years at Yale, and my embarrassed exit from the profession. I would need to look up my former classmates and professors; review the research on women's performance in STEM fields ("STEM" being the shorthand of the day for "science, technology, engineering, and mathematics"); return to my high school in upstate New York, and then to my college campus in New Haven, to see what had or had not changed in the decades since I had studied there; and talk to the young female students who surrounded me daily on the campus of the less elitist but equally rigorous public university where I teach creative writing.

The statistics I found depressed me. Even today, only one-fifth of all physics PhDs in this country are awarded to women; only about half of those degrees go to native-born Americans; and only about 14 percent of all physics professors in the United States are female. The numbers of black and Hispanic scientists are even lower—in a typical year, only about a dozen African Americans and twenty Latinos of either gender receive doctoral degrees in physics. But the reasons for the shortages of minority scientists don't seem mysterious: blacks, Hispanics, and Native Americans often attend schools that leave them too far behind ever to catch up in science or math, and the effect of prejudice at every stage of their careers has been well documented. What I couldn't figure out was what might still be keeping white middle-class women out of careers that offer so many opportunities in terms of job prospects, prestige, intellectual stimulation, and income.

Even in the 1970s, the sexism I experienced was rarely obvious. I grew up in a privileged, loving home with few barriers that might prevent a bright, confident young woman from succeeding in whatever field she took it in her head to enter. All this led me to suspect that the reasons for the scarcity of female physicists must be subtle, and those reasons must lie buried in the psyches of the women who loved science and math but never completed their degrees or, like me, earned their degrees but left their fields. Which didn't mean such factors were indescribable. What seemed necessary was an approach that combined a rigorous dissection of my own experience, coupled with interviews with other women and an analysis of the studies that psychologists and sociologists had carried out. By trying to understand why I didn't become a physicist, I hoped to gain insights into why so many young women still fail to go on in science and math in the numbers their presence in high school classrooms and their scores on standardized tests predict.

What I discovered shocked me. Although more young women major in physics at Yale than when I attended school there, those young women told me stories of the sexism they had encountered in junior high and high school that seemed even more troubling than what I had experienced: complaints about being belittled and teased by their classmates and teachers, worries about being perceived as unfeminine or uncool. If anything, the recent sexualization of American culture—our obsession with princesses and pornography, our romanticization of marriage, housekeeping, cooking, and motherhood—has created even more intense pressures on women who pursue a career that isn't perceived as typically female or whose requirements aren't compatible with the demands of traditional suburban life. The same forces that caused me to feel isolated and unsure of myself at Yale continue to hem in young women today, acting like an invisible electrified field to discourage all but the thickest skinned from following their passion for science, a phenomenon that turns out to be less true in other countries, where women are perceived as being equally capable in science and math as men.

Girls are still steered away from or allowed to drop advanced courses in science and math. This is especially insidious given that girls who are good at science and math tend to be equally adept in the humanities; they find themselves praised for their talents as readers and writers and nudged toward so-called "people" fields, while boys who are gifted in science tend to be more narrow in their abilities and receive far more reinforcement for their high scores in calculus, physics, or computer science. Young women still enter college less prepared in math and science than their male classmates, only to be shocked by their failing grades and the rigorous demands of their majors. Rather than receive tutoring or support, such students are quickly and intentionally weeded out.

Perhaps my most significant finding is that female science majors need far more encouragement than men, even as their instructors perceive any need for praise as a sign that a student lacks the seriousness or commitment to succeed in research. The prevailing ethos tends to be, *Anyone who needs to be encouraged shouldn't be*, a philosophy that might seem equitable if not that most women are actively discouraged from pursuing careers in science—or difficult, demanding careers in anything. Most science and math instructors believe they are being evenhanded in their refusal to encourage anyone, not understanding that any white male who grows up in this country already receives encouragement for his ambitions, if only in the form of the prevailing image of scientists as white and male. More than that, studies have documented that professors of both genders do respond more positively to their white male students; for instance, they are more likely to agree to meet a prospective student if the request comes from someone whose name is stereotypically white and male.

Given my own experience, I wasn't surprised that psychologists have documented the power of "the stereotype effect"—the distraction of being the only woman or black person (or only anything) in the room, the self-fulfilling prophesy of fearing that you will do poorly on an exam because women or black people supposedly aren't good at whatever subject is being tested—although I was startled how easily such detrimental effects can be mitigated by informing the test taker that the stereotypes aren't true. Similarly, I was gratified to learn that a woman's ineptness in a laboratory usually is a result of her culturally induced perception that she doesn't belong in such a setting and her lack of familiarity with the apparatus; even if small biological differences in spatial reasoning between the genders do exist, these can be evened out by a brief course designed to make women and minorities feel comfortable in a lab.

Although I didn't go on to graduate school, my interviews with women who did confirmed my fear that I would have been turned off by the competitive, aggressive atmosphere, as well as by the attitude that scientists must focus so obsessively on their research that they have no time for anything resembling a normal life. Even in college, the expectation that I give up reading and writing and playing tennis, as well as friendships with other women and romantic fulfillment with a man, made me wary of applying to graduate programs in theoretical physics.

And yet, even if I had fulfilled my dream of attaining a PhD in cosmology from Princeton, the pervasive bias against women in science, combined with my lack of confidence in my abilities, might have prevented me from rising to prominence, or even remaining in the field. By far the most upsetting finding I discovered is that subtle biases against female scientists still prevail, even in supposedly female-friendly disciplines such as biology. As a study published in the fall of 2012 demonstrates, scientists at every level—old or young, female or male, in physics, biology, chemistry, and engineering—still view men as more hirable, more competent, more worthy of mentoring, and worthier of higher salaries than women with exactly the same qualifications. No wonder I felt underappreciated by my professors, even as my classmates insinuated I was receiving whatever opportunities I was receiving because my university was trying to fulfill its affirmative action quotas.

As other studies demonstrate, biases against female scientists don't stop just because a researcher lands a job. Even the most talented, well-trained women grow frustrated by their failure to publish articles or win grants they might have been awarded if a male researcher's name had graced the byline; by the smaller labs, lower salaries, and fewer resources they receive compared to the male researchers next door; by the higher standards they must meet to achieve promotions; and by the less desirable teaching assignments they tend to be assigned. Add to that unequal burdens of childcare and housework and the difficulties that arise from the need to find specialized jobs in the same city for both a husband and a wife and you are left with female scientists who grow so exhausted they "choose" to stay home to raise their kids or leave technical fields for less demanding or more welcoming positions. This reinforces the belief that women don't have what it takes to succeed at the highest levels of science, or that they prefer to deploy their talents in so-called "people" fields such as medicine, teaching, or law, or caring for their children, or, in my case, writing.

In theoretical physics and computer science, arrogance and aggression can be the norm, with a premium placed on around-the-clock work broken only by recreational pursuits generally favored by adolescent males. Such clubhouse camaraderie often includes misogynistic jokes and hazing directed at the one or two women in the room. A female scientist's attempt to discuss her discomfort generally will be met by silence, the assertion that such antics are in fun, or the attitude that any discussion of one's feelings is out of place in a realm where logic and facts are the accepted currency.

Deviating from the norm in even one respect makes success in the STEM fields difficult. Deviating in more than one respect—being female *and* poor, being female *and* gay, being black or Hispanic or Native American *and* attending a less-than-stellar high school—makes success nearly impossible. Perhaps the greatest reason for resistance to making physics and math more welcoming is the false assumption that if anyone needs a few extra semesters to make up for a weak beginning, or assistance with childcare, or a flexible schedule to tend to the demands of life outside the lab, that person doesn't belong in science. Most scientists believe that changing anything about their field would be equivalent to dumbing down the discipline and sabotaging America's traditional excellence in pure and applied research, even as the current system discourages women and minorities whose talents and creativity might exceed those of mediocre white male scientists who receive far better educations and far more encouragement from their parents, teachers, classmates, and society. The system condemns itself to a narrow range of outlooks and approaches, even though science profits from the broadest possible spectrum of perspectives.

And yet, the very fact that most of the obstacles preventing women from succeeding in science are cultural and psychological rather than legal or institutional gives me hope that change can come by making otherwise well-intentioned people aware of the subtle biases and illogical assumptions that drive women from the field. If women truly were biologically incapable of performing well in science and math, whether in the middle of the curve or at the very highest end, we might need to give up the fantasy of increasing their numbers. But as we will see in later chapters, a wide array of studies demonstrates this isn't the case. Instead of trying to elevate a young woman's IQ, all we need to do is elevate her confidence; provide her with images of female scientists and toys that nurture more than her desire to dress up like a princess; encourage her not to drop AP calculus; change the problems in her textbooks so they don't presume an interest in football or war; allow her to appreciate the joys of designing a computer game that doesn't involve blowing up people's heads, in a room that isn't populated solely by farting, burping, breast-ogling young men. Rather than struggle to pass new laws, we can work to convince male scientists to let go of the belief that rudeness and aggression are essential for success. Even if we can't overcome the anti-intellectualism that has so long pervaded American culture, we can attempt to counter the equation of an aptitude for science or math with nerdy unpopularity, a prejudice that strikes most cruelly at adolescent girls—precisely at a time when an exaggerated concern for social status can ensure they will never be able to pursue careers in science.

One of the most gratifying aspects of writing this book has been witnessing how quickly a woman can recognize the ways in which her own sexism, lack of confidence, and craving for praise have been preventing her from achieving all she might achieve and causing her to undervalue the talents of other female scientists. I have experienced firsthand the efficacy of workshops that make faculty aware of the ways in which a dearth of women and minorities on search committees can perpetuate the dearth of female and minority candidates for new positions, or the ways in which professors of both genders sabotage their female students by writing shorter, weaker letters of recommendation than they would write for their male protégés. But no one will be motivated to make such changes or attend such workshops if he or she believes that bringing more women and minorities into the field will produce inferior research. No one will trade a system that seeks to weed out as many young scientists as possible for a policy that instills a love of science and math in those who otherwise might never discover such a love, a policy that builds the confidence of those whom society has sapped of confidence and helps those who have received an inferior education to overcome their early deficit, unless he or she becomes convinced that the young people we lose to other disciplines might have become superior scientists. As is true with most campaigns for reform, nothing will change until we can put faces and names to the undervalued, until we can document what we have lost, and continue to lose, by our ignorance or denial.

In many ways, my own story has a happy ending. I love being a writer, and I doubt the world lost its chance to understand the origins of the universe because I didn't study theoretical physics at Princeton. But my memories of adolescence and young adulthood vibrate with disappointment. When I decided to write this book, I went down to the basement and opened my father's footlocker from World War II, into which I had stuffed all my papers and books from Yale. As I sifted through its contents, I became paralyzed with pain. Could I really have been the girl who wrote such difficult equations—and understood what they meant? When was the last time I experienced the thrill of grasping a fundamental truth about the universe? How could I have doubted I wasn't talented enough to go on for a PhD? How could so much hard work have come to nothing? Opening that trunk was like finding the souvenirs of a sojourn in a foreign land where I had spent four years laboring to make the desert bloom, all while learning a language so difficult that I could no longer recall the alphabet. Yet here was a pressed hibiscus from that garden, and what seemed to be an entire epic poem composed in that foreign tongue, written in a hand I barely recognized as my own.

Odder still: in the years since I left that country, most of us have come to live there, if only in the sense that we have grown so addicted to new technologies. Yet even as geeks like Bill Gates, Steve Jobs, and Mark Zuckerberg have become hipper, wealthier, and more prominent in our society, the gender of those geeks remains the same (the number of women in computer science has *dropped* in the past few years). I can't help but think that the scarcity of female physicists is related not only to the obstacles I faced in the 1970s, when I struggled to obtain my degree in physics, but also to my reluctance to write this book. Even as women have published memoirs about being sexually abused, addicted to crack or alcohol, too thin or too fat, too beautiful, too rich, too poor, too manic or too depressed, no woman seems willing to confess in print to loving science or math too much.

The reasons so few girls consider careers in those subjects, and the reasons so many women who passionately love science and math walk away from their dreams, as I walked away from mine, are rooted in the psychology of each individual woman and the culture in which all of us live. Studies can go only so far in documenting why anyone might choose to pursue this career above some other—or decide to stay home and raise her kids. Such decisions might have their origins in our biology, but that doesn't mean they aren't also heavily influenced by the expectations we pick up from our parents, siblings, friends, classmates, teachers, guidance counselors, partners, and potential lovers; from the toys we are—or are not—given to play with as children; from the images we absorb from movies, television shows, novels, and advertisements. How could I not have been affected by the fact that every scientist I saw growing up and every voice I heard narrate a documentary or museum exhibit was male? From the time I got glasses in second grade, I repeatedly heard the warning that no man would make a pass at a girl who wore them, the larger implication being (and I understood this even in second grade) that glasses not only would mar my beauty but signify that I was "smart."

Like many smart girls, I made astonishingly stupid choices. What doesn't show up in most studies of female scientists is how strongly their psyches are shaped by the traumas of junior high. Even women who grow up to be feisty, successful feminists spend much of their adolescence obsessing about their appearance, romance, sex, and their popularity with female friends. If the basketball players their own age won't date them (and in junior high and high school, even the smart, socially awkward boys moon over the conventionally popular or pretty girls), girls may dumb themselves down, hide or repress their interest in classes or activities their peers deem nerdy. They may develop crushes on their teachers and other older men, who don't see them as threatening and are all too happy to reciprocate their affections. A boy might pursue a subject because he respects the man who teaches it, but unless he is gay, he won't fall in love with that teacher, as so many young women do.

In junior high and high school, when science and math seem challenging and girls begin to feel isolated or ostracized within such courses, the social pressures I try to capture in the opening chapters of this book loom larger than one would think. A nerdy boy might be as undatable as a nerdy girl, but at least the boy can dream that he will grow up to become a physicist whose Nobel Prize will attract a mate, or a technocrat like Zuckerberg or Jobs whose fortune will allow him to marry well. By contrast, the nerdy girl is baffled by the realization that the more obviously successful she becomes, the more likely she is to find herself sitting at home on a Friday night. Not to mention that the men who teach English, drama, or art in high school may be easier to fall in love with than men who teach physics, chemistry, or mathematics, if only because the material they cover in class is more likely to cast a romantic aura about their heads.

In writing this book, I came to see that my long-suppressed feelings about my body and my femininity, my teenage desire to think myself popular, sexy, cool, and my anxieties about my ability to attract a man played an embarrassingly large role in my decision to give up physics. Even my Orthodox Jewish upbringing revealed itself as a factor: examining my experience at Yale and comparing it to the experiences of other women, I discovered that anyone who already feels out of place because she is too dark skinned, poor, working class, gay, or even, in my case, too uncultured and too Jewish, might not risk further isolation by majoring in a subject in which she feels doubly an outsider, especially at an exclusive institution such as Yale.

To understand why so many male scientists shrug off their female students' decisions to drop their courses, I needed to examine the ways in which my parents and teachers steered me toward the humanities and other "people" subjects, then used my eventual choice to become a writer as evidence that girls innately prefer such subjects. I needed to figure out why young women earn praise for editing the newspaper or cheering on the cheerleading squad—even, in rare cases, debating on the debate team—but confront indifference or hostility if they express an interest in signing up for a computer club, a science fair, or a math olympiad. I needed to remember how bored I was at school, how few teachers seemed willing or able to keep me occupied, how much more emphasis was placed on my failure to be ladylike and well behaved than my passion for learning. I needed to explore the ways in which an outing to a world's fair or a natural history museum can ignite a child's curiosity about a subject to which her parents or teachers might otherwise never expose her. I needed to reteach myself algebra, geometry, trigonometry, calculus, and physics so I could understand the ways in which the problems and illustrations in a textbook can shape a young woman's enjoyment of a subject and distort her perception as to whether or not that subject is meant for her, or only for human beings who are intimate with footballs, cars, bombs, and guns.

So entangled was my decision to leave physics with other aspects of my life that I even needed to examine such seemingly irrelevant factors as my love of tennis—and the reasons I felt so uneasy about winning a match. The more I talked to other women, the more I came to see that the ability of female students to compete in sports not only provides an outlet for their frustrations and allows them to retain their physical and mental health while struggling to succeed in graduate school, it also helps them to develop the ability to tough out challenges, to bear other people's rudeness and aggression, to get dirty and sweaty without fear they will be less attractive.

To tell my story, I needed to move backward and forward in time, compare my memories to those of others, take measure of my own perceptions from other observers' points of view. Just as Georges Seurat was able to create an impression of Parisians enjoying a Sunday afternoon on the River Seine by painting thousands of tiny colored dots, I want to convey what it felt like to be an intelligent, ambitious young woman growing up in the late sixties and early seventies, and why even today so few women and minorities go on in science. A more apt metaphor might come from calculus, in which the area beneath a curve can be derived by slicing the irregular shape into rectangles whose uppermost borders don't quite fit the curve, then dividing the shape into ever-finer rectangles until their numbers approach infinity and the area beneath the curve approaches the sum of those infinitely many, infinitely narrow strips. Maybe if I provide enough fragments of my experience, the sum of those glimpses will create an impression of a girl who wanted to launch herself at the cosmos, only to fall back to Earth because she barely understood her own heart and mind, let alone the world she lived in.

PART I

Leaving Liberty

CHAPTER ONE

A Different Kind of Math

Most scientists believe that anyone who is destined to join their ranks will demonstrate a love for the subject nearly from birth and nothing will discourage such a child from realizing his abilities. That I gave up physics to become a writer proves I couldn't have been serious in the first place.

And yet, a more accurate predictor of success, whether in science or the humanities, might be an intense curiosity, a refusal to take for granted any of the miracles most children are taught to accept as banal reality. Maybe because I have been blessed and cursed with disturbingly vivid dreams—my nighttime adventures are so convincing that even now, I often awake convinced that I have flown—I became obsessed at any early age with the workings of my own mind. Once, when I was three or four, I told my older sister smugly: "I'm talking but you can't hear."

"Oh," she said. "What you're doing is called *thinking*. Everyone can think."

"Everyone?" I was disappointed. But it struck me as astonishing that each of us could talk inside our heads without anyone on the outside hearing.

What amazed me was so few grown-ups shared my amazement at what was going on inside our heads. On the first day of kindergarten,

I amused myself by drawing the numeral four with a slanty top instead of the recommended three-sided square. I even convinced my friend Denise to draw her four with a slanty top, which resulted in the teacher scolding me for leading my friend astray. Worse, my teacher's response convinced me that I would be better off not asking the question I wanted to ask: namely, who had decided what symbol should represent four pennies in the first place?

By the first week of first grade, I was so bored I sat squirming in my chair as our teacher turned the pages of the giant Dick and Jane primer she had propped on an easel. The teacher's name, Mrs. Prettyman, evoked a woman who was a man, or a man who was pretty instead of handsome. I often found myself worrying about whether I was a girl pretending to be a boy or a boy pretending to be a girl, so this caused me no small alarm. I was even more horrified to see our teacher tug my friend Steven up the aisle, pull down his pants, and spank him. When she suggested I curb my own tendencies toward disturbing my neighbors by sitting quietly at my desk and "writing a story," I took out a sheet of paper and printed the words my sister had taught me the day before: *I'm looking over a four-leaf clover that I overlooked before . . .*

Capturing a story in faint gray pencil-marks on the page was as exciting as capturing a ghost and rendering it visible to others. But when I showed my creation to Mrs. Prettyman, she shook her head and said, "That's not a story, that's a song." Puzzled, I went back to my desk and copied down "Cinderella." Mrs. Prettyman explained in an even less patient voice that I should *make up my own story.* Then I understood: Everyone was entitled to make up her own story! You pulled one story from your head, and another story popped up in its place, like tissues from a box. That I could create something out of nothing caused me to grow lightheaded. I loved creating something out of nothing. I loved talking inside my head. I loved marking words on a page so other people could read them and hear my voice inside of theirs.

My enjoyment of school didn't really sour until third grade. We were studying Indians that year, for no reason I could see except that good

Indians were defined by their ability to walk down the hall in two straight lines without speaking to their partners. One afternoon, we were making headbands on which to glue the feathers we had won for good behavior when a stranger knocked at our classroom door. The edges of my headband kept sticking to my fingers. Not that I had earned any feathers to paste on that headband anyway. I looked up and saw my teacher crook her finger to summon me to the hall.

"This is Mr. Spiro," she said. "He wants to talk to you in his office." Then she went back in and shut the door.

I had been causing a lot of trouble. I still missed my second-grade teacher, Mrs. Hoos, who had joked she was going to bring in her laundry to keep me occupied but allowed me to sit at my desk, reading book after book, as long as I read quietly. The principal's wife, Mrs. Neff, was made of starchier, sterner stuff. By God, if we were reading aloud, paragraph by painful paragraph, I was going to sit there with my book open to the appropriate page and not read a word ahead.

My only pleasure came from whatever proximity I could finagle to either of the boys I had a crush on. My neighbor Harry would grow up to be class president and, despite his lack of height, captain of the basketball team. But the real object of my affection that year was Eric, who had appeared out of nowhere. In gym, waiting for the teacher to put on a record so we could square dance, he admitted he had been skipped ahead because second grade bored him. Furious that this gangly boy had been allowed to skip second grade while I was being bored to death in third, I punched Eric in the stomach. Then, as he gasped for air, I realized I had found an ally against the stultifying enemy that plagued us both, and I asked my new friend to dance.

As it turned out, Eric considered third-grade math to be even less riveting than second-grade math, and he and Harry and I persuaded Mrs. Neff to allow us to work in a workbook designed to teach us to multiply and divide while our classmates practiced addition and subtraction. But whenever I whispered a joke to Harry or poked Eric in the spine, I lost the privilege of working in my workbook, which made me misbehave even more. On this cycle went until multiplication seemed a meal I would never get to eat because I was so weakened by hunger I couldn't reach the plate. I tormented Mrs. Neff, and she tormented me, until we each wished the other gone. And because she was the teacher, she had the power to make her wish come true.

Warily, I followed Mr. Spiro to the school's top floor, then down a short, dark passage, where we climbed to an oddly shaped room with a slanted ceiling. Mr. Spiro settled behind his desk and asked me to sit beside it. Bushy black hair exploded from his head like the lines a cartoonist draws to indicate that a character is confused or drunk. *Mr. Spiral*, I remember thinking, which is how, half a century later, I can still recall his name. His heavy eyebrows, gigantic nose, and thickly thatched mustache seemed connected to his glasses. His suit was white with red stripes, like the boxes movie popcorn came in, and he wore a red bowtie. This was 1964. I had never seen a man in a shirt that wasn't white or a suit that wasn't dark or a tie that called attention to itself, and I felt the dread of any child who has been selected from the audience by a clown.

"Would you like some Oreos?" he asked, sliding a packet across the desk.

I can still see those cookies, so chocolaty rich and round. The red ribbon around their cellophane cocoon just waited to be unzipped. But a voice in my head warned me to be careful. It was as if the clown had motioned me to sniff the carnation fastened to his lapel. "How do I know these cookies aren't poisoned?" I asked.

The bushy black brows shot up. "Do you really think I would poison you?"

"You're a stranger," I said. "How do I know you wouldn't?"

Stunned, he changed the subject. "So!" he exclaimed. "I hear that you want to be an authoress!"

If he had asked if I wanted to be an *author*, I would have told him yes. But I had never heard that word, *authoress*, and it seemed dangerous as a snake. *Authoress*, it hissed, like *adulteress*, a word I had encountered in a book and didn't understand, except to know that I didn't want to be one. "Who told you that?" I demanded.

"Why, your teacher, Mrs. Neff."

"That's because she hates me. She probably told you a lot of other lies, but those aren't true either."

He scribbled something on a pad, and even a child of eight knows anything an adult writes about you on a pad can't do you much good. Outside, a pair of pigeons bobbled back and forth across the ledge like seedy vaudevillians trading a joke (this was the Borscht Belt, after all), then shuckling back across the stage.

Mr. Spiro finished writing. "I hear you are a bright little girl, and I would like to give you some tests designed to demonstrate if you are smart enough to skip a grade."

Skip a grade? Like Eric? My heart began to race. *Bring on the tests!* I thought, expecting a mimeographed sheet of addition and subtraction problems. Instead, Mr. Spiro pulled out a flip-book and showed me drawings, asking me to describe what was missing from each. But how could a person know what was missing from a picture if there was nothing to compare it to? Did every house have a chimney? Was the daisy *missing* a petal, or had someone merely plucked it?

After we finished the flip-book, Mr. Spiro brought out a board fitted with colored shapes. He would flash a pattern and ask me to reproduce it. But as vivid as my memory is for people and events, I have a terrible time remembering patterns and facts. Had the purple triangle been positioned above or below the line? Had the rectangle been blue or green? And who had decreed that playing with colored shapes should determine if a child was ready to skip a grade? It wasn't enough to be smart; you needed to be smart in the ways grown-ups wanted you to be smart.

Finally, Mr. Spiro presented me with a test like the ones I was accustomed to taking. "If you want to buy three pencils, and one pencil costs four cents, how much would it cost to buy all three?"

I was about to add three and four when I realized that a different kind of math was called for. I could have added four three times, but something about being so bored I had misbehaved—which denied me the right to work in my workbook, which denied me the chance to learn to multiply, which seemed to be preventing me from going on to fourth grade, where I might be less bored and less tempted to misbehave—made me cry out, "That isn't fair!"

But when Mr. Spiro asked *what* wasn't fair, I couldn't put my grievance into words. We sat in silent stalemate until a pigeon flew in the window. It flapped around our heads and beat its wings against the walls. But this struck me as no more unusual than anything else that had happened. For all I knew, Mr. Spiro had trained the pigeon to fly in on cue and test some aspect of my psychology I would need in order to pass fourth grade. The safest response seemed no response at all.

Mr. Spiro, on the other hand, leapt up on his desk and began waving his arms and shouting. Imagine sitting in a chair looking up at a full-grown man in a red-and-white-striped suit who is standing on his desk flailing at a pigeon. I might have understood if he had been flailing at a bee, but what harm could a pigeon do?

The pigeon found a crevice and disappeared inside—my final glimpse of its bobbing rear end is the image I still see whenever I hear the word *pigeonholed*. Mr. Spiro climbed down from his desk and asked why I hadn't been more upset. All these years later, I remember what I said. "Why should I be upset? This isn't my office. I'm not the one who needs to clean up after it."

A disheveled Mr. Spiro led me back to class. Later, he told my mother I wasn't "mature" or "well socialized" enough to skip a grade. Even then, I dimly recognized I was being held to a standard Eric had not been held to—a year younger, he was far less mature or well socialized than I was. The experience left me more resentful. The following year, my teacher grew so impatient with my incessant chattering that in front of the entire class, she said, "Eileen, has anyone ever told you how obnoxious you are?" *Obnoxious*, I repeated, delighted and appalled by the toxicity of the word. *Obnoxious*, was I? Fine. I shunned the company of other girls and hung around with the roughest boys, who were even more obnoxious than I was. I still did well on tests—what was I supposed to do, pretend I didn't know how to add (or multiply)? But I refused to act the part of the well-mannered little lady the grown-ups wanted me to play.

CHAPTER TWO

Science Fair

Like any child, what I craved was magic. Writing was a form of magic, but so was science, the difference being that scientific magic was much harder to get to work and you needed a grown-up to help. When called on to construct a project for your third-grade science class, you couldn't simply rely on a pencil, a sheet of paper, and your imagination. At home, you might think you had built a working model of the solar system. But in the shoebox you brought to school, all anyone could see was a wad of aluminum foil, a pink Spaldeen, a Ping-Pong ball, a cherry, a roll of cellophane tape, a spool of thread, and a wire hanger.

I had an easier time with Legos. With Legos, you could fit the pegs on one block into the holes on another block with a satisfying click, then add a pulley, a hook, four walls, and voilà, you had a castle with a working drawbridge or, if you added wheels, a pickup truck or a tank. I begged my parents for an Erector Set, but that's where they drew the line. Unlike Legos, which came in colorful plastic shapes, the pieces of an Erector Set were stiff metal strips a child combined by screwing them together. Only boys were allowed to own Erector Sets. Playing with my friend Jeffrey and *his* Erector Set in his shadowy garage on a sunny day felt forbidden and electric.

At some point, though, Jeff and I must have grown tired of the Erector Set, because we went outside and knelt in the driveway, where we used his new magnifying glass to funnel rays from the sun onto a scrap of paper. I will never forget the bright white circle of light, like the gaze of the Evil Eye my parents so feared would direct its attention on any child whose relatives praised her for being too pretty, too strong, too smart, or the gunpowdery scent of the paper as it started to smoke, or the pounding of my heart as the center of the beam burst into flame.

I don't remember learning much science in elementary school. Year after year, we propped a celery stalk in red or blue ink and watched the fluid creep up the vegetable's pale green veins. Or we stuffed blotting paper in a jar, wet the cardboard, then slid lima beans between the paper and the glass and took notes as the beans shed their skins and unfurled their cottony roots. Enticing the beans to perform these acts in public struck me as illicit. But our teachers asked us to carry out the same two experiments every year, just as they taught us, over and over, how to add and subtract and divide a pie.

Most of what I learned about science, I learned at home. Someone had given my brother a chemistry set for his bar mitzvah, and, four years his junior, I surreptitiously claimed that gift as mine. The white metal box opened to form a three-sided cabinet of wonders. Nestled within its walls were containers filled with colored powders; grainy strips that changed from pink to blue—or blue to pink—when you dipped them in a solution whose acidity or alkalinity you wished to test; metal spatulas; rubber stoppers drilled with holes you could fit with plastic tubes; a scale whose cunning brass weights were shaped like pillbox hats; beakers, flasks, and tongs; a pipette with a rubber bulb; a glass lamp such as Ali Baba might have owned (you filled the lamp with alcohol, screwed on a lid whose wick you lit with matches you had stolen from the kitchen, then placed the burner beneath a tripod upon which you set whatever concoction you were brewing that day); and, most mysterious of all, a black plastic eyepiece you held to your eye so you could glimpse the flashing particles emitted by the lump of radioactive ore inside.

Best of all, the set came with a manual that allowed you to carry out actual experiments. If you mixed this chemical with some other, you ended up with a vial of disappearing ink. Obey the instructions, and a yellowish puff of bad-egg stench wafted from a test tube like a curse from an infant's mouth. I was especially intrigued by crystals. Dip a string in a solution of sugar water and you had rock candy. Swab Epsom salts on a tumbler, and the glass would be etched with a Jack Frost pattern as fancy as the crystal in your mother's cabinet.

I kept waiting for my brother to figure out I was using his chemistry set and hit me. But he didn't seem interested in the gift. Nor did he yell at me for borrowing his microscope, which I used to study flakes of skin, or a fragment of a fly or moth. When the town's water supply became contaminated with bright red rotifers, I captured a few in an eyedropper, squirted them on a slide, and watched as they darted here and there, powered by the miniscule wheels that spun at the tips of their seahorse tails. I refused to drink the water. But I reversed my earlier refusal to take a bath and settled in the tub amid my newfound friends, who zipped playfully around my toes as I enjoyed my childhood version of swimming with the dolphins.

My brother also owned a dry-cell battery. Don't ask me why. Dry-cell batteries were simply what boys in those days owned. He taught me to twist one end of a wire around one node, then attach the other end to a buzzer or a bell or a bulb, then attach another wire to the other screw on the buzzer or bell or bulb, then attach the other end of that wire to the second node on the battery, which caused the buzzer to buzz or the hammer to ping the bell or the miniature bulb to glow. I didn't know what an electron was, or how a wire full of them might cause a buzzer to buzz. But I *wanted* to understand.

My brother's first science fair project was the classic egg to chick. We lived in chicken-farming country, so he had no trouble finding fertilized eggs. These he set in an incubator in the basement, and every afternoon, he would select a partially gestated egg, tweeze a window in the shell, and draw a poster of what he saw. I would follow him downstairs and stare at that day's embryo, which stared back with its huge black eye, the red pinprick of a heart already beating, until, at the end of three weeks, a nearly developed chick lay wetly in its dish, one damp wing curled across its face as if to shield it from my spying.

I had been trying not to think where babies came from, but my brother's project cracked the shell I had constructed around my ignorance. My curiosity burst out and started to squawk, and no matter how hard I tried, I couldn't stuff it back inside or quiet it. Joking, my brother demanded which came first, the chicken or the egg, and I couldn't stop thinking about the answer. His posters detailed which parts of the chick developed on which day, but how did a cell know it was destined to become a beak, a wing, a brain? Around this same time, *Life* came out with its famous photos of a fetus inside a womb, and it occurred to me that a person might achieve the fame I craved by figuring out how a living creature grew. I don't remember if my brother's project won a prize. But that project changed my life.

As a middle schooler, I loved everything about the science fairs in our town, especially being greeted by a nodding, grinning robot I thought of as Mr. Lightbulb. Mr. Lightbulb and I had previously been introduced at the General Electric Pavilion at the 1964 World's Fair in Flushing, New York, when I was eight, and I couldn't have been more excited if Frosty the Snowman had honored us with his presence.

My parents had taken us to Flushing Meadows twice—they enjoyed the fair as much as we did—and nothing in my childhood excited me half as much. At the GE Pavilion, where Mr. Lightbulb waved you in, the Carousel of Progress stunned me with the idea that we in the audience might keep revolving forever around a stage on which each Audio-Animatronics family enjoyed devices whose function and design were ever more fantastic and impossible to predict.

What those exhibits instilled was awe. You entered the Bell System Pavilion believing the world had always been the way it was. Then you rode in your armchair—sideways—while a resonant voice murmured through metal earphones the size of grapefruits that Man had progressed from banging on drums to inventing the written word, which he learned to chisel in stone, then inscribe on parchment with a feather pen, then print on paper with a press. Next came the telegraph, the telephone, and then the videophone, which allowed you to see the person to whom you spoke (back home, we still placed our calls by means of an operator who demanded in a nasal whine, "*Num*-ber, please!"). The message you kept receiving was that human beings had evolved from nothing, and then they invented *this*.

And oh, that Voice. The Voice of Science. Reason. Progress. I couldn't get enough of that Voice of Awe. I took it home as a souvenir, not from the Bell System Pavilion, but from a building whose red clamshell roof I assumed must be a flying saucer, not realizing until years later it was the umbrella for the Travelers Insurance Company. No moving armchairs here. You walked from one glassed-in display to the next and peered in voyeuristically at the hairy, half-naked people—maybe, if you stood there long enough, the boy in the ragged skins would wander over and press his fingertips to your own. I refused to leave until my parents bought me the red 45-rpm recording that would allow me to replay the Voice of Awe at home. I slid that disc on the phallic spindle of my family's record player and listened to the Voice describe the dioramas. And each time I listened, I experienced the thrill anew. Something had evolved from nothing. Didn't anyone understand what the Voice of Awe was saying—that it was a miracle we were *us*?

But what the future often reveals isn't more future; what the future reveals is the past we have lost. When I google the Travelers Insurance Pavilion on the Internet (an invention predicted nowhere at the fair), I discover someone has uploaded that long-vanished disc, with photos of the dioramas. Clicking on the booklet's cover, I find myself standing beside my parents, listening as the trumpets blare to herald the "unforgettable journey" I am about to take. I see those naked, furry men thrusting spears into a giant warthog. *These are your ancestors of a million and a half years ago. They are ape-like and covered with coarse hair, but they can think, they can learn, and the simple tools they are using to kill these warthogs eventually will take Man to the stars.*

And here she is, the hairy-breasted woman who has been crouching in the cave inside my brain for nearly five decades, warming her hands at a fire as her shaggy child gnaws on a bone and the father of the child stands silhouetted against the snow, returning from the hunt. The next few dioramas rekindle other memories: The Origin of Art; The Beginning of Agriculture; The Grandeur of Rome; and, with a dread that comes over me like the flu, a jumble of corpses as *the death wagon rolls through cobbled streets*, people coughing and convulsing in alleys, rats with gleaming eyes scurrying around their legs, every feverish detail as vivid as if the germs of this plague have been lying dormant in my unconscious to reinfect me now. Then, the final diorama: Man's Leap to the Stars, in which Man comes to understand that *even as he frees himself from the planet which has held him captive for almost two billion years, he must never forget that where there is love, he must protect.*

That exhibit became my psyche. The resonant male Voice on that recording echoed inside my skull for the decades that followed. And yet, I can't help but note that the Voice intones the words *Man* and *Mankind* more than forty times; the male pronoun and its variations—*he, him,* and *his*—almost an equal number; the word *woman,* not once; and the female pronoun and its variations, zero. The only female figures in the dioramas are the mother in that cave feeding her genderless child, and—with her husband's arm wrapped protectively around her waist—the wife of Daniel Boone.

Every spring, the sixth graders in our town traveled to the Museum of Natural History—we lived only ninety miles from Manhattan, but most of my classmates had never been there. After we caught a glimpse of the Oz-like skyline, we tunneled beneath the Hudson to reemerge on streets shadowed by skyscrapers, each of which contained more inhabitants than our town. The museum's dioramas were even more impressive than their counterparts at the world's fair. The jawbones and tibias might have been shards of Adam's thigh, that's how deeply moved I was. In the planetarium, I leaned back and listened as the Voice of Awe transported me to the incomprehensible Nothingness before time and space began.[1]

Some of my classmates fell asleep. Others used the Nothingness to make out. I felt as lonely as a girl who, on a pilgrimage to the Mother Church, decides to become a nun, only to glimpse her friends necking in the pew beside her. All of it, Science and God and Sex, got mixed up in my brain. I wanted to be a biologist and a magician, a rabbi and an archaeologist, a prophetess and an author.

Until the following year, when I watched Neil Armstrong frolic on the moon and decided to become an astronaut.

So serious was this ambition that I conditioned myself by doing chin-ups and spinning on a swing as a prelude to getting spun in that nausea-inducing bucket they had at NASA. When my brother discovered me carrying out such preparations, he took pleasure in informing me that astronauts needed twenty-twenty vision so their glasses wouldn't go floating off in space. This put a crimp in my plans, but I figured I could perform exercises to improve my eyesight. At least he hadn't said women couldn't be astronauts. I could improve my vision but not change my sex.

Around that same time, my brother chose as his science fair project the theory of relativity. The images of rockets flying so fast the rulers inside them shrank and passengers aged at a slower rate than their siblings who stayed home were as enticing as a brochure for Wonderland. Best of all, it was a Wonderland I could reach, if only I studied each poster's captions and asked my brother to repeat his explanations of what each one meant.

There were no science fairs in elementary school, but in sixth grade we took standardized tests called the Iowas, which were used to select the Smartest Boy and the Smartest Girl, each of whom would receive a dictionary embossed with his or her name. Those titles—Smartest

---

1. On a recent return to the planetarium, I was delighted to find that the male Voice of Awe had been replaced by Whoopi Goldberg's.

Boy and Smartest Girl—made sense to everyone, as if intelligence were a contest in which the sexes couldn't compete on equal terms. Worse, I started to get the message that competing with boys in any way would win me no friends of either gender. I overheard my mother chide my sister for beating a boy at Ping-Pong and wondered how anyone could *pretend* to lose at Ping-Pong. Not only was the title of Smartest Girl inferior to the Smartest Boy, even to be called onstage to accept that dictionary was to risk teasing and unpopularity.

I couldn't ask anyone at school for guidance. At best, the adults treated me with neglect. More often, they resented me for asking too many questions. Worst of all, like my sixth-grade teacher, Mr. F., they praised me to the skies.

I knew I was in for a terrible time that year, but my parents were friends with Mr. F.'s parents and they couldn't ask that I be reassigned. My anxiety escalated when I followed the directions to my room, turned down a dead-end hall, and saw the alcove that led to what had once been Mr. Spiro's office. Apparently, Mr. F. had persuaded the administration to let him board up the windows and use the belfry as a darkroom. Or he had simply gone ahead and done it.

Mr. F. seemed even more bored by school than I was. He would put his feet on his desk, tell us how desperate he was to work full time as a photographer, then describe the Hasselblad camera he was saving to buy. If a male classmate misbehaved, Mr. F. yanked him into the darkroom and we heard terrible bangs and crashes. But nothing I did, no matter how unruly, earned me a reprimand. My status as the Smartest Girl, coupled with my complete lack of social grace, would have made me a pariah anyway. But Mr. F. brought that fate upon me sooner by handing back an exam on which I had scored an A, asking me to stand, and demanding that my classmates be more like me.

Complaining about being praised is like complaining about being pretty. Even then I knew it was better to be me than Pablo Rodriguez, whose parents were migrant workers, or the Buck brothers, Phil and Gregory, who seemed to get punished for no other reason than they were large and male and black. But if I was so angry and unhappy, it defies me to imagine how much angrier and unhappier kids like Pablo and Phil and Gregory must have been.

. . .

To escape, I read science fiction. My favorite was the Mushroom Planet series, which chronicled the adventures of two boys, David and Chuck, who discover an ad in their local paper:

WANTED: A small space ship about eight feet long, built by a boy, or by two boys, between the ages of eight and eleven. . . . Please bring your ship *as soon as possible* to Mr. Tyco M. Bass, 5 Thallo Street, Pacific Grove, California.

In the first book, David and Chuck rescue the inhabitants of a planet called Basidium by sniffing the sulphury stench of the plants that provide their food and deducing that the nutrient the Basidium-ites lack can be supplied by hard-boiled eggs. In the next installment (published in 1956, the year I was published, too), the author, Eleanor Cameron, had the foresight to subject her heroes to a close encounter with something she called "a hole in space, a hole in nothingness."

I didn't mind that the characters in the books were boys. A woman had written the series, which made her more powerful than its heroes. What stopped my fantasies cold was that David and Chuck were so much handier at building things than I was. After cutting a sheet of "plastiglass" to form their spaceship's windshield, they bolted it down and sealed it with a rubber sealant. "Boy, what a beautiful job!" Chuck sighed proudly. True, Mr. Bass was guiding their hands. But my experience with snippers and sealants was so limited, not even a wizard could have helped me to build a spaceship.

On TV, my favorite show was *Star Trek*, which I watched with my brother on the small TV he had purchased with tips earned from waiting tables at a local hotel. Usually, if he found me in his room, he threw me out. But he allowed me in to watch *Star Trek*, especially if I agreed to walk on his back, which ached from carrying all those trays. "Live long and prosper," we would say, flashing the secret Vulcan sign. My brother was dispassionate and aloof, and my desire to win his approval became displaced onto my desire to be respected and loved by the chief science officer of the *Starship Enterprise*.

Best of all, Mr. Spock had female colleagues. Not only Lieutenant Uhura, glorified receptionist that she was, or even Nurse Chapel, who shared my crush on Mr. Spock, but various ensigns, doctors, "yeomen," and psychiatrists, many of whom were accorded the respect of being allowed to die as they did their jobs. I would go to bed early so I could write and star in my own episodes. There we would be, with even Mr. Spock stumped as to how to save us, and I would volunteer a solution. One pointy Vulcan eyebrow would lift as he recognized our common superiority. Later, on some planet whose orbit was being disrupted by a Romulan force field, causing papier-mâché rocks to tumble about our heads, I would use my lightning reflexes to save his life. He would carry me to the ship and wait impassively as Bones waved his instruments across my chest and snarled at his Vulcan crewmate, "Damn you, Spock, the poor girl saved your life! Can't you at least show some concern for whether she pulls through?"

I knew how much Spock cared, even if he found it sloppily human to express his feelings. I could only hope for sex every seven years, when Vulcans were in heat. But what I wanted more than sex was the opportunity to fly to distant sectors of the universe, discover new forms of life, and boldly go where no woman had gone before.

I might have continued reading about outer space forever if I hadn't developed a more pressing need for advice about how to survive on Earth. When my former friends began to shun me, when they invited me to parties that didn't exist solely for the pleasure of hiding behind a bush and watching me confront an empty backyard, when they invented an entire new language for the purpose of mocking me, when they told me the boys on the basketball team would never date girls who did better in school than they did, I turned to novels for reassurance that clever, resourceful young women like Jane Eyre, Jo March, or, God help me, the heroines of *The Fountainhead* and *Atlas Shrugged* found lovers who were even smarter and more powerful than they were.

Then again, it's not as if anyone encouraged me to continue reading science fiction. One afternoon, I wandered into Krug's Stationery, the only establishment in town that sold books, and picked out a trilogy by Isaac Asimov and a paperback called *One, Two, Three . . . Infinity* by George Gamow. When I carried these to the register, the owner said, "Oh no, dear, those books are *boys'* books. The girls' books are in the back."

Puzzled, I went to see what books she meant, only to discover that the "girls' books" consisted of a rack of romance novels.

CHAPTER THREE

Science Unfair

All through elementary school I looked forward to junior high, believing I wouldn't be forced to learn to add, subtract, and multiply for the billionth time. But seventh-grade math was no more challenging than fifth or sixth. The first week of term, I filed into the auditorium with ninety other seventh graders to take an exam whose purpose remained unclear until my friends Eric and Jeff vanished from my science and math classes. In those days, fundamentalist Christians kept quiet about their beliefs, so I had never heard anyone talk about the Rapture. But I can only compare my distress at learning Eric and Jeff had been skipped ahead to what a Christian might experience if her peers had been wafted to heaven while she bided her time in a squalid waiting room with a few outdated *Reader's Digest*s to keep her occupied.

I was even more upset when Eric revealed he had overheard our principal, Mr. Van Slyke, say I hadn't been skipped ahead because girls never completed programs in science or math. Eric told me that he tried to argue, but Mr. Van wouldn't hear a word.

Eric remembers none of this. The person whose casual good deed saves someone else's life rarely recalls the incident with the same clarity as the person whose life he saved. The irony is that after our junior year in high school, Eric never took another course in math. He went on to Harvard, earned his degree from Harvard Law, and helped to found one of the most prestigious litigation firms in New York. He lives in Manhattan, where he and his wife have been able to send their three children to the finest private schools. But even there, he says, the girls in his sons' classes don't find it cool to do well in math and science.

Jeff is more like me: he remembers everything about his childhood. Apparently, he wondered why I hadn't been skipped ahead. "It didn't make any sense. They could easily have spread the minimal resources they were spending to three of us instead of two." Jeff credits his decision to major in physics at Cornell to the physics class he took his junior year in Liberty. The physics teacher, Mr. Yates, entered him in a program at Bell Labs, and the two of them traveled to New Jersey for a weekend to take a tour. "Lasers were new, computers were new—we got to see all that," Jeff tells me now. "After that, I was very into cosmology and astronomy. That's what got me into science as an undergrad."

As Jeff tells me about his years at Microsoft, where the miniscule number of women engineers depressed him, I flash to a memory of Eric and Jeff in the AV room playing with the school's primitive computer. "That old thing?" Jeff scoffs. "It was a programmable HP calculator. It used reverse Polish notation. It must have weighed a hundred pounds, and the computer part went back, like, two feet from the screen."

Still, I can't help but think how much more comfortable I would have been with computers in college if I had been invited to hack around with that primitive HP in high school. I ask Jeff if he ever entered the science fairs. Yeah, he says. He did a project on electricity based on an idea he got from *Popular Science*. His uncle, an engineer, helped him with the soldering. Together, they rigged a contraption that involved two running streams of water, coffee cans, copper hoops, and a coat hanger; the hoops built up an electric charge that caused a spark to leap the gap, a process I don't quite follow because I am lost in envy at the idea of an adult teaching me to solder.

After Eric spilled the beans that I hadn't been allowed to skip ahead because girls never finish programs in science or math, I sent my mother to Parents' Night to complain. As I recall, she confronted Mr. Van, the principal, and Ed Wolff, the chair of the Math Department, both of whom seemed surprised I would care. One or both of the men assured her it wasn't a good idea to narrow a child's horizons too early. Besides, a girl who got skipped ahead in math might find her social life had been destroyed.

My mother wasn't in the habit of challenging rules set down by men. She hadn't defied her own father and gone to college. How could she champion her daughter's right to take courses in science and math the male students were allowed to take? What's less obvious is why I didn't champion my own right to take those courses. Why didn't I simply show up for the same algebra sessions as Jeff and Eric? What would their tutor, Mr. Gallagher, have done, hauled me from my seat and carried me back to my seventh-grade math class?

For all I considered myself a rebel, I became viscerally ill if I broke a rule. I didn't smoke or get high or drink. I was secretly relieved that my parents prohibited me from attending Woodstock, which took place a few miles down the road. Who would want to sleep in the mud, dance naked in the rain, or eat slop ladled out by dirty, tattooed men who belonged to a commune called the Pig Farm? The nightly newscasts from Vietnam upset me terribly, but I was even more anxious about the protests disrupting the nation's colleges. *Sure*, I thought, *by the time I get out of Liberty, the universities will all be closed.*

My opinions of the women's movement were even more contradictory. My friends and I spoke out in support of the Equal Rights Amendment and proclaimed our willingness to call boys on the phone or ask them to dance (not that we exercised either prerogative). But the women's movement seemed to mean women ended up spending more time with other women, in something called "consciousness-raising groups," and the last thing I wanted was to spend more time with women. If women ran the world, society would be less competitive. But I loved competing. How else could I prove to the brilliant, powerful men who ruled the world that I was as smart and strong as they were?

Then again, maybe I didn't stand up to my jailers because I realized how quickly I would have been overpowered. Mr. Van was a dour, no-nonsense vet with the hollow eyes and chiseled Nordic features of a Haldeman or an Ehrlichman. And Ed Wolff, who taught every math class I ever took, except for trigonometry with Mr. Gallagher, was tall and squarely built. Neither was a man even the toughest kids cared to cross. So I accepted my jail was real and the best I could do was to be a sullen and rebellious prisoner, a female Cool Hand Luke. I shuffled to math class late, then sat in the back row and cracked snide comments. I followed my teacher's lectures only to catch him in an error. Listening to my objection that some proof was wrong, he would rock back on his heels, fold his arms across his chest, stroke his heavy, square chin, and shake his head in what seemed an impersonation of his slightly more famous look-alike, Ed Sullivan, with that impresario's mordant wit and impatience with any act that wasn't up to snuff. "Eileen," he would drawl, "you're not nearly as smart as you think you are."

Once, after I said something more disrespectful than usual, he asked why I couldn't be more ladylike and well behaved like my sister, Joan. That filled me with rage . . . and broke my heart. At sleepover parties, while my friends mooned about their crushes on Ed Wolff, I sat desolate, knowing how much he loathed me, knowing how much I loathed him, yet wishing he liked me better, because secretly, I had a crush on Ed Wolff, too.

It never occurred to me not to care what Ed Wolff thought. Instead of banding with the other girls to change the system, I needed to prove I was the worthiest to succeed. Unlike all the other women who hadn't finished their programs in science and math, I was serious and smart enough to finish mine.

Like me, many of my female classmates thought the battle for equal rights had already been won. And yet, our social studies teacher, Mr. Burke, warned us that boys would be happy to liberate us of our virginity, but they would marry girls who just said no. A few of my classmates traveled to other towns and found doctors to prescribe the Pill, but their mothers always found the plastic clamshells they had squirreled away in their underwear drawers. I never could be sure who was going all the way with their boyfriends; when we talked about sex, we giggled and employed euphemisms that could have meant anything, or, in my case, nothing.

One day, I overheard the girls at my lunch table laughing because I carried my books against my hip like a boy instead of crushed to my chest as they did. So that was why no one asked me to make out at parties! It wasn't only that I was smart; I didn't know how to be a girl. I took out my yellow planner and designed a rigorous campaign to rectify the situation. First, I studied the behavior of the one girl I wanted most to be, the irony being that what I admired most about Denise was how utterly satisfied she was to be herself. Despite taking the fewest pains with her appearance, she was the prettiest of us all. She radiated inner calm, but she had a sly smile and a wicked laugh and knew the value of a well-placed, deflating barb. I couldn't put any of this into words, so at the top of my list of rules for self-improvement, I merely wrote: TRY TO BE MORE DENISEY, followed by:

DON'T ASK QUESTIONS IN CLASS

NEVER LET ANYONE KNOW WHAT YOU GOT ON A TEST

DON'T OFFER TO HELP THE GUYS WITH THEIR MATH HOMEWORK

TALK MORE SOFTLY.

Then, as if I hadn't already silenced myself enough, I added: ALWAYS THINK BEFORE YOU SPEAK. Now, thinking before you speak can be a useful rule. But with so much second-guessing going on inside your head, how can you concentrate on anything going on outside it?

Worst of all, I developed the habit of making self-deprecating jokes about myself, as if to say, "See? I can put myself down more effectively than any of you could ever do." At first, I didn't believe what I was saying. After a while, I did.

Even as I was trying to remake myself in the image of a fifties bobbysoxer, my mother had decided to liberate herself and go to college. Years earlier, her teachers had urged her to apply, but her father didn't see the need because she would only end up married. Then he died, and she went to work to support her mother. All I knew about her job was that her boss kept chasing her around his desk, trying to pinch her bottom. She was bright and hyper-competent, but she lacked confidence and so funneled her energy into keeping our house obsessively neat and ironing my father's underwear.

Finally, in the late sixties, she started taking courses at the community college. My father had always been the parent whose absence haunted me, but suddenly the equation changed and my mother's was the time I craved. Why had I assumed my father was the smarter of my parents? My mother earned all As and was invited to give the valedictory address at her graduation. Responding to the turmoil of the times, she quoted a poem about a falcon circling in a gyre and some rough beast slouching toward Bethlehem, and even though she spoke in that prissy voice I hated, I realized I was proud of her and would need to go to the library and find the poem.

After that, she became a different person. Happy and brave, she drove an hour each way over the steep and winding Minnewaska Trail to finish her degree at SUNY New Paltz. Occasionally, she would allow me to skip school and accompany her. Once, she took me on a bird-banding expedition with a female biology professor. Another time, she burst into our kitchen exulting at her success at dissecting a fetal pig—the younger students had been too squeamish, but she had called upon her experience reaching into the cavities of kosher chickens and yanking out their guts. Left to herself, my mother might have majored in biology. But my father convinced her that studying English would be easier, and, in the end, she got a job teaching third grade.

In middle school, my science classes were memorable for little but the grainy black-and-white movies in which meteorologists in white short-sleeve shirts raised their voices to be heard above the drone of an engine while pointing out the window of an airplane at the nimbus, stratus, cirrus, or cumulus clouds below. For Earth Science, I had as my teacher a kind, bumbling man named Harry Wach, who also ran the youth services at our synagogue. Mr. Wach was difficult to dislike (his nickname was "Wacky Wach"). But as the weeks dragged on and we sat through more and more films, I became more and more unruly.

Finally, Mr. Wach announced that although we were too young to enter the real science fair, he was requiring us to do a project. Desperate to prove an injustice had been perpetuated by leaving such a talented young scientist in the care of a teacher named Wacky Wach, I searched our *World Book Encyclopedia* for a project sure to dazzle. I didn't care what the topic was, as long as I could perform the experiments at home, the way a novice cook searches for a recipe that doesn't require *ricing* or *poaching* or ingredients she can't obtain from her local grocer. Under *E*, I came upon *Electroplating*. The encyclopedia was written with a thirteen-year-old reader in mind, so I understood the basics. If you took an object made of one metal, connected it to the negative pole of a battery, and immersed it in a solution containing a positively charged salt of another metal, you could cause the object to be plated with ions from the solution, although why anyone would want to do this I hadn't a clue.

I hooked up my brother's dry-cell battery to a spoon. In a glass baking-dish I constructed a rickety apparatus involving a quarter and a key, some aluminum foil, water, kosher salt, and nails from my father's workshop. A few bubbles of what I hoped to be chlorine gas rose from the nail, but I couldn't coax the tiniest flake of metal to deposit itself on the spoon, and when the due date came, I had nothing to show except a lengthy write-up of my experiment. Even so, I was startled when Wacky Wach handed back my composition book with a B and expressed his disappointment that I hadn't come up with a better project.

Today, when I google *electroplating*, what pops up is a frustrated query from a ninth-grade boy named Andre G., who worries that "avoiding acid mists from the electrolyte" might be a problem and plating an object with chromium or zinc might release poisonous cyanide gas, followed by a reply from a man named Ted Mooney, whose mission seems to be offering guidance to any student attempting to demonstrate the wonders of electroplating. In the condescending tone engineers use when describing an industrial process to a layperson, Mr. Mooney recommends employing a copper penny as a cathode, a zinc rod as an anode, and a solution of zinc dissolved in vinegar and water to plate the penny with the zinc. As a child, I might have been able to find a penny, vinegar, and water. But when Mr. Mooney assures his audience that "zinc anodes are available from boating stores," I can't help but shake my head. Even if I had found a boating store, pedaled there on my bike, and persuaded the owner to sell me a zinc anode, how could I have proceeded to the final step, implied by Mr. Mooney's cryptic statement that "with a hacksaw a teacher can cut many slices from one anode"? The image of Wacky Wach hacking at a hunk of zinc includes blood and flying fingers. Nor would I have been brave enough to follow Mr. Mooney's suggestion that I slice open a store-bought battery, my parents having warned me that to do so would result in acid exploding in my face. (Then again, they also said that if I cut open a golf ball, the pressurized core would explode and blind me.)

Maybe now, as a grown-up, I could follow Mr. Mooney's instructions. But as a kid, I wouldn't have had the slightest hope of accomplishing anything he advised. I'm lucky that whatever gas I managed to release wasn't actually chlorine or I would have burned my esophagus. I'm glad I didn't electrocute myself with that dry-cell battery. All of which goes to prove that long before the Mr. Mooneys of the world were able to promulgate their advice via the web, a child needed more than a copy of the *World Book Encyclopedia* to pull off a project like the one I attempted. Even the brightest kid needs a sympathetic grown-up to drive her to a boating store and hack apart a battery. When she shows up with nothing but a cooking pan and a spoon, a wad of aluminum foil and a battery, her teacher can't simply say, *I'm disappointed. You get a B.*

In ninth grade, when I was eligible to enter the real science fair, I again consulted our encyclopedia. This time, under *H*, I found *Hydroponics*. Because the project relied on knowledge I had gleaned from all those bean-growing experiments, I trusted I might succeed. I convinced the pharmacist to sell me the nutrients I needed to feed my plants. After determining what "vermiculite" was, I carted home a sack from the gardening store. (The vermiculite was required to prevent the tomatoes from toppling over, although I couldn't understand the point of growing tomatoes in vermiculite when I could have grown them in regular soil in the first place.) I found chicken wire at Frankel's Hardware and cut squares to lay across the rubber pans I bought at Woolworth's. Every afternoon, I ran upstairs and measured nutrients with the scale from my brother's chemistry set, then mixed that day's solutions in my mother's enamel pot, my hands shaking because my parents had warned me that anything I spilled would come flooding through the ceiling.

Watching my radishes grow wasn't as eerie as peeking inside the eggs my brother had tweezed to reveal those beating hearts. But I took pleasure in drawing diagrams of sprouting seeds and recording the precise composition of each nutrient solution and the corresponding rates of growth. The first radish I harvested nestled in my palm like a sunburned Thumbelina; I could barely bring myself to take my mother's knife to the bulb and slice a dozen round white slices.

Best of all: I got to buy three creamy sheets of oak tag, a pack of stencils, and a box of Magic Markers, whose very name implied powers beyond what an ordinary writing implement might accomplish. What a satisfying squeak those markers made! Using stencils, even a klutz like me could draw a line so straight you couldn't help but think what else you might accomplish if provided the proper tools. In a rare display of solidarity, my brother showed me how to frame each poster, tape wire hangers to the backs, then use masking tape to connect all three to form a triptych. When I despaired of conveying my vegetables to the fair, my mother grudgingly agreed to drive.

More surprising: my chief competitor turned out to be Linda K., whose project involved finding the optimal growing conditions for her mother's African violets. I hated Linda K., even as I envied her power to judge who among us did or did not wear stylish clothes, who was or wasn't cool enough to date a member of the basketball team, and who would—or would never—learn to dance (as Linda put it, "Give up, Eileen, you have no soul"). How could anyone have predicted that instead of pronouncing the science fair a retarded waste of time, Linda would take it as her goal to win? Although what chance did she have for that? Her project involved discovering the ideal growing conditions for a *houseplant* rather than the radishes and tomatoes that might feed the inhabitants of a country blessed with copious supplies of Epsom salts, water, and vermiculite, but utterly lacking in dirt.

Neither Eric nor Jeff entered the fair that year, which meant any victory I achieved was hollow. Nor do I remember a word of praise from my parents when I announced I had won first prize. But I was ecstatic to have pulled off a project that didn't dissolve into a mess of chicken wire and vermiculite but produced an edible crop of fresh— if antiseptic—vegetables that I arranged neatly on a plate and served the judges.

By tenth grade, the science fair was no longer cool and no one entered. Still, I decided to undertake an extra-credit project for biology class. A book called *The Population Bomb* had just been published, warning of an explosion in the number of people on the planet, millions of whom would starve, and talk of the population bomb was everywhere, the way talk of global warming is everywhere now. I had never cared much about biology. Imagine sticking a Post-it to a wet wall and you will understand what it was like for me to try to memorize the organelles of a cell or the steps in the nitrogen fixation cycle. But my project on population growth required no rote learning. I could go to the biology room after school, and my teacher, Mr. Nickou, would teach me to mix batches of agar-agar, then pour the warm gelatin into petri dishes, where it hardened to form a surface as pristine as the skating rink at Grossinger's after the Zamboni had cleared the ice. I sterilized a wire loop, dipped it in bacteria, inoculated each dish, then set the stack of plates in a dark, warm drawer as I charted each colony's growth against the piece of graph paper I had taped to the back. I studied my poor, ignorant subjects as they ate and reproduced until they overfilled their puny world (oh, I had read my Malthus!), ran out of food, and suffered a Dark Ages of the kind I had witnessed at the Travelers Insurance Pavilion, the wagons rumbling through the streets stacked with the corpses of starved bacteria, until the survivors vanquished their grief and began to repopulate their dish—two survivors begetting four, four begetting eight—although my subjects never did seem to learn and again would outstrip their limits.

My biology teacher seemed excited by my success, which I couldn't have achieved if not for his willingness to stay after school and help. Most teachers were either loved or loathed, but Mr. Nickou was neither. Students kept a tally of how many times he cleared his throat, a rate that shot up astronomically when he reached the unit on human reproduction. But I didn't rate my teachers according to their quirks. I liked anyone who wanted to pass along the mystery of his subject. Which, in his own dry way, Mr. Nickou did. The organs inside the pickled grasshopper staked to our dissecting tray might resemble shavings from a pencil, but this seemingly insignificant insect had evolved a digestive system so brilliantly economical that instead of excreting urine, which would squander its single teardrop of essential fluid, the grasshopper's Malpighian tubules filtered the waste from its blood, allowing the water to be reabsorbed into its gut, after which the insect excreted dry crystals of uric acid. And how could a person not find it interesting that earthworms were hermaphrodites and reproduced by having intercourse with themselves?

The highlight of the year was the day we received our frogs. We pictured our teacher, knobby-kneed in shorts, scooping bullfrogs from a pond, then plopping them in a bucket. Did he pith each frog himself? (The word *pith* sent a jolt along my spine, as if my own vertebrae had received the needle.) No matter how the amphibian on our tray had met its end, there the rubbery creature stretched. One moment, you were staring at a frog whose belly had been sliced; the next, you were swept up in the understanding that all this complexity was improbable to the point of miracle, and your own guts even more so. I became fascinated by the human brain, the way trillions of neurons and synapses intersected to create a self that was capable of wondering how evolution had produced the very brain that was wondering how it evolved. When Mr. Nickou told us a biologist had zapped a solution of amino acids with electricity to simulate the lightning that must have struck the amino-rich scum floating atop the murky soup of the earth's primordial seas, setting off a chain of biochemical events that led to the first protozoans, I imagined those first living cells as the iridescent blobs on my grandmother's chicken soup, then tried to wrap my mind around the fact that at some point those blobs had started *living*. The thrill was a high I could re-create any time, especially if I needed to numb the pain of feeling so poorly adapted to the environment in which I needed to survive at school.

In chemistry, my teacher was Wilmer Sipple, whose bald head and bespectacled face reminded me of a honeydew melon. (Years later, when I saw a Muppet scientist whose name was Bunsen Honeydew, I wondered if everyone had a chemistry teacher whose nearly featureless head reminded him of a melon.) For all that, he was a sweet, good-hearted man, and he explained chemical reactions clearly. The atoms of one element lacked a certain number of electrons to complete their outer shells, while the atoms of some other element possessed extra electrons in their outer shells, so the two elements could share electrons and feel complete. Pluses attracted minuses. Everything in the universe was made of these elements. Chemistry made simple, Sipple sense.

And I looked forward to the labs. How could history or English compete with the pleasure of fitting goggles around your head, connecting the hose of a burner to a jet, then flicking the handles of a striker so the flint would ignite the gas? Other days, Mr. Sipple provided us with palettes of solutions that reminded me of the watercolor sets I had loved as a kid, challenging us to identify the liquid in each well. The experiments were easy to perform. And our teacher's corny but good-natured puns made the time pass quickly.

I even looked forward to the semester's end, when we would take the Regents Exam in chemistry. No one at our school had achieved a perfect score, and Mr. Sipple renewed his promise that he would buy a surf-and-turf dinner at the Red Barn restaurant for any student who pulled off this feat. I hadn't a clue what a surf-and-turf dinner was, but I welcomed the chance to prove to Mr. Van and Mr. Wolff that I was worthy to succeed in science. By now, my classmates and I had struck a deal: I wouldn't try to date anyone on the basketball team, in return for which they would treat me as a mascot whose achievements reflected well on them. And so, when Mr. Sipple announced my score, everyone chorused in jubilation, "Way to go, Eileen! Now we'll be remembered as the class that won Mr. Sipple's bet!" Although they couldn't help but add, "Ha, ha. Now you're going to have to spend an entire dinner with Wilmer Sipple!"

This reminder gave me pause. Mr. Sipple was one of the nerdiest teachers in our school. No doubt his wife would be nerdy, too. What on earth would we find to talk about?

The fated hour came. Mr. Sipple and his wife drove up to my house, and for the first time, I wondered if the mere existence of a competition was sufficient reason to try to win it. We traveled to the restaurant in silence; there, the waitress handed us menus, which allowed me to determine that *turf* meant steak and *surf* meant lobster, a disconcerting revelation given I had never eaten such a blatantly nonkosher meal. Sensing my hesitation, but interpreting it as reluctance to order the most expensive item, Mr. Sipple cried, "A bet's a bet!" and ordered the entrée for me (although I noticed he ordered the much cheaper flounder for Mrs. Sipple and himself).

Happily, I discovered that although the gristly kosher beef my mother served at home repulsed me, the London broil at the Red Barn melted in my mouth, and the tender flesh a person could obtain by disassembling a lobster (its carapace gave way with the same satisfying *pop* as the grasshopper I had dissected) was so succulent I developed a lifelong taste for *treyfe*. Mr. Sipple and his wife turned out to be adept conversationalists, and I didn't need to struggle to keep up my end—Mrs. Sipple was an artist, and she had a lot more on the ball than most women I knew. A year later, when I entered Yale, I felt confident enough to sign up for advanced physical chemistry. I wish someone had warned me that the stopcock on a burette needed to be closed before you poured in the acid. But I can't blame my affable Muppet of a teacher for failing to warn me that when hydrochloric acid gets dribbled on a woman's stockings, the nylon goes up in smoke.

I liked my physics teacher, too. The only blight on our relationship was that I'd had his wife for seventh-grade English and we had been at war all year. She was a severe woman who lectured us on the difference between *Scottish* and *Scotch*, showed us the tartan plaid that symbolized her clan, and, when I spoke out of turn, said, "Eileen, I have some muzzles at home that I use for my Scotties, and I am going to bring them to class to use on you!" But Mr. Yates didn't share his wife's desire to muzzle me, and I did well enough that he invited me to join the engineering club. I was more interested in thinking about space and time than building radios. And, I am ashamed to say, the boys in the radio club held less allure than the tantalizingly out-of-reach members of the basketball team. Maybe if I had joined his club, Mr. Yates would have invited me to visit Bell Labs in New Jersey, the way he invited Jeff. Then again, he might have thought better than to take a weekend trip with a teenage girl.

CHAPTER FOUR

Advanced Placement

Every girl who ends up majoring in English instead of physics does so because she has a teacher like Barry Talkington. Math and science teachers can be handsome and debonair. But classes in math and science don't lend themselves to discussing love as readily as a lesson on *Romeo and Juliet*.

By eighth grade, when I tried out for the debate team, I already had a crush on Barry (even as students, we called him Barry). He was in his midtwenties, with long, scraggly hair, a bumpy aquiline nose, lively dark eyes, and a receding hairline that made him resemble the young Will Shakespeare. Not only did Barry judge me to be verbally dexterous enough to join the debate team; the qualities that made most other adults dislike me—my sense of humor, my hunger to be the best at everything, my curiosity about the world—caused Barry to like me more.

Eric tried out, too, and he and I got paired, as we always did, the Smartest Girl and the Smartest Boy. I wasn't the mastermind of our team. But I could think on my feet, which made me good at cross-examination. And since I held so few political beliefs of my own, I had no problem arguing the affirmative or the negative of any proposition.

By junior year, Eric and I had become one of the strongest debate teams in the country. Paired with two seniors, we beat schools twenty times our size and far better funded. Competing at such a high level was difficult for a boy, but I needed to be extra careful not to alienate the nuns and priests who judged for the Catholic teams, or the Southern coaches, who hated to see a girl speak quickly or act aggressively. It was like competing in the steer-wrestling event while wearing high heels and a skirt and making sure not to hurt the steer's feelings.

Still, debate provided me with the opportunity to feel desirable. The male members of our debate team ("master debaters," they called themselves) were witty and smart. I dated Eric, until he dropped me for a younger woman—a cheerleader, of all things. (I can't say I blame him. The implication of our always being paired was that no one would date the Smartest Boy except the Smartest Girl, and vice versa.) Rather than replace him with another boy on our team, I used our weekend trips to enjoy the attentions of debaters from all-boy schools. Once, an Irish kid from Boston who looked like Jack Kennedy's seedy younger brother came up and whispered, "You have nice legs," a fact I found more illuminating than anything that pertained to that year's topic.

Mostly, I mooned around our coach. It would take another book to describe how much I loved Barry, or all the things he taught me—how to drive a car with a stick shift (I blew out the clutch on the high school van), how to appreciate the stories of Flannery O'Connor and, I am embarrassed to say, the philosophy of Ayn Rand. (Neither of us could help but be drawn to a world in which our superior qualities would be recognized by a virile superman who would carry us off and ravish us.) Today, when I look at a yearbook candid of Barry, arms raised, knees bent, performing an ironic pas de deux with a younger female member of the debate team ("Eat your heart out Ginger Rogers!!!" the caption reads), I marvel it never occurred to me that he might be gay. Not only did I not know any openly homosexual classmates in high school, I didn't meet any openly homosexual people of either gender at Yale. As happens with so many women who fall in love with men who aren't able to explain why they don't find them sexually exciting, Barry's inability to admit he loved men nearly destroyed us both.

Not until the end of my senior year did our romance blossom into anything the least bit physical. But in high school, romance was all I needed. That, and the electricity of sitting next to Barry in the van, inhaling the aphrodisiacal scent of cigarettes and beer embedded in his leather jacket, knowing a handsome older man found me to be attractive and predicted that once I got out of Liberty, I would lead an exciting and successful life.

I would have spent every day that summer hanging around Barry's cottage, but like most kids in those days, I was expected to work. My grandparents had sold our hotel, so I couldn't find employment there. After one stint changing diapers as a mother's helper, and another waiting tables at Howard Johnson's, I tried out for a secretarial job at the company that insured the few remaining resorts in the Catskills. I appreciated that I no longer needed to scurry around HoJo's bearing fried clams and banana splits. But I resented not being allowed to go out "in the field" with the male adjustors. Instead, I filled out claim forms and typed up the adjustors' notes of their investigations, growing increasingly phobic about engaging in any of the activities the hotel guests engaged in, although my greater fear was I might end up like the older secretaries, their hair sculpted into beehives above painted-on eyebrows, mouths as red as the waxy lips my friends and I used to buy at the gas station across from our elementary school.

As often as I could, I visited the adjustors in their smoke-clogged office in the back, convinced that because they were men, their conversation must be more enlightening, even though all they did was regale each other with descriptions of the female guests they had shtupped or the boobies on the women walking on the street below. Once, they showed me photos of a topless softball game they had played against the strippers from the club across from the racetrack in Monticello.

Most nights, I drove out to Barry's house. A single male teacher hanging around with a female student wasn't the scandal it would be today. But it wasn't typical behavior. Years later, I asked my mother why she allowed me to hang around with Barry, and she said, "I figured he was either a homosexual, in which case you weren't in any danger, or he wasn't, in which case there were worse things than you falling in love with a nice, educated man and marrying him. I guess I thought that if you ended up marrying Barry, you would come back to Liberty, teach English, and settle down, and it would be nice to have you nearby."

That fall—the autumn of my senior year—I would get up in the morning, eat breakfast, then saunter down the street so my home-room teacher could check off my name. I attended physics class, then walked back up the street and spent the rest of the day in my parents' attic, preparing for the four AP examinations I was determined to take that spring. Whenever I finished a chapter in calculus or biology, I rewarded myself by fixing a snack; in photos from that year, a roll of extra flesh swells above the waist of my plaid bell-bottom pants.

At three, when everyone else was leaving school, I walked back to help Barry and Eric train the junior members of the debate team. In earlier years, I had edited the school newspaper and participated in the usual roster of activities that ambitious students join to prove how civic-minded they are. But now I devoted all my spare time to preparing for my exams and figuring out which colleges to apply to.

I had no one to advise me. Our guidance counselor was famous for telling kids they needn't bother to apply to college. But if he thought you had a prayer of getting into an Ivy League school, he paid you so much attention he might have been sitting atop your head. The day our class filed into the gym to take our SATs (an exam for which no one in those days prepared), he positioned himself behind me and watched as I colored in the ovals for the first few questions. Furious that after eighteen years of neglect I should bring the slightest distinction to anyone in my hometown now, I began coloring in the ovals beneath the answer C, over and over. The counselor let out a gasp. I turned and sliced my finger across my neck, then jabbed it toward the opposite corner of the gym. After he reluctantly complied, I wasted another few minutes erasing all the Cs and started over.

Once I got my scores, I began combing through the guide to colleges my sister had purchased six years earlier. The book was so out of date I couldn't tell which universities remained all male and which had gone co-ed. Was I supposed to apply to Barnard . . . or Columbia? To Pembroke . . . or Brown? I had done well enough on my SATs that I received an invitation to apply to MIT, and I added Princeton to my list because Princeton was, well, Princeton. I chose the University of Rochester because my sister and brother had gone there. An older boy named Dean Silverman, whose curly brown hair and fine features distracted me from prayers at services, had startled our community by earning a place at Yale, this being the first time any young man from Liberty—certainly any Jew—had accomplished this feat, so I penciled Yale on my list, with a question mark that meant I didn't know if it admitted women.

When the applications came, I used my mother's manual typewriter to fill them out. I could have shown Barry what I wrote, but I was too embarrassed. The one piece of advice I took—my counselor had recommended I lie and say I wanted to be a doctor rather than a physicist so the admissions committee wouldn't think me odd—I still regret. I devoted more time to typing my essay in the box without making any errors than revising what I wrote. Even so, I spoiled my application to MIT; not wanting to alert the admissions office to my incompetence by requesting another form, I retyped the essay on a clean sheet of paper, then cut it to fit the box and taped it to the application, which is how I mailed it in.

I applied to Yale early decision, and in December, when the acceptance arrived, I was overcome with joy. But the decision wasn't binding, and I didn't receive a dime in financial aid. We weren't poor, but my father charged a small-town dentist's rates. He and my mother were supporting both my grandmothers in nursing homes and had just put my two older siblings through college. So when MIT and the University of Rochester said I wouldn't need to pay anything to attend their schools, and Harvard, Princeton, Cornell, and Brown offered me admission but, like Yale, gave no financial aid, I persuaded myself I would be happy at MIT. How could I not be happy at a school where everyone studied physics and men so outnumbered women that I would surely have every boy on campus longing to ask me out?

Then, on a debate trip to Boston, Barry drove me to MIT. "Jesus," he said. "Are you sure you want to go to a school where the buildings have numbers instead of names?" Lost, we dead-ended in an alley surrounded by engineering labs. When Barry asked if I wanted to get out and look around, I said, "Uh, no," then tried to figure out how I was going to break the news to my father that I wanted to give up a scholarship to MIT.

If I wasn't going to MIT, what college was I going to attend? Harvard was where the senior member of our debate team had gone, and where Eric was going now. If I went with him, I would spend the next four years mouthing the words he and the other boys gave me to mouth. Another member of our team had gone to Princeton, but according to this young man, if you went to Princeton, you needed to belong to something called an eating club, and I'd had enough of snobby cliques in high school. Which left only Brown and Yale.

But Brown was the most radical of the Ivies, and I hated the idea of not getting grades. I was more impressed when I took a tour of Yale and the guide informed us that classes there were taught by full professors and students were required to take nine courses a year instead of eight. The sky was a bright, crisp blue, and the Gothic buildings and plush courtyards shone at their aristocratic best. The professor who taught the introductory physics class was bearded, young, and dark, with shoulders so broad they covered half the equations he was writing on the board. When he turned to face the auditorium, his eyes were such a radiant blue they seemed to be powered by some radioactive element. Sprinting across the stage, he wove the blackboard with a lacework of symbols I couldn't unravel and told physics jokes whose punch lines flew above my head. But after the lecture, as I watched him make his way up the aisle while students clustered in his path, trying to ask him questions, I decided I wanted nothing more than to be one of those students. And so, when I got home, I sent in my reply to Yale that I was coming.

The end of the school year came. I missed earning a perfect score on my AP exam in English because I misspelled *boulevard* as *bullevard*. But where would I have seen the word? Liberty had avenues and streets, but not a boulevard. I got fives in calculus, chemistry, and biology and was named valedictorian of our class. Everyone had expected Eric and me to share the stage. But the salutatorian was a girl named Sharon, who had studied harder and gotten better grades than the Smartest Boy.

Sharon was more modest and well behaved than I was, and I didn't really get to know her until senior year, when I was astonished to learn she planned to major in physics at SUNY Binghamton. I broached the idea that instead of writing two speeches, we should collaborate on a comic dialogue about being female and smart in a school where those two adjectives spelled your doom. We worked on it for days, laughing and shaking our heads at our outrageous bravery. Our triumph was tempered by a summons from Mr. Van. I was sure he would refuse to let us deliver a speech in which we skewered everyone in the school, including him. But for whatever reason— maybe he was tired of fighting me and looked forward to my departure—he granted his permission.

The harsher critic turned out to be my mother, who threatened to boycott graduation. "Fine!" I said. "It's not as if you ever cared about anything I ever did anyway!"

Sharon and I marched down the aisle and took our place beside our classmates. I looked out and saw my parents sitting in the last row, presumably so they could slip out if they grew embarrassed. After the invocation and a speech by the class president—my old friend Harry—it was time for Sharon and me to deliver our routine. Two microphones had been provided. Sharon bent one up a few inches; I bent the other down.

"We thought we would give a speech about women's lib," Sharon informed the audience.

"Don't worry, though," I said. "We're not going to speak about *that* subject. Two years in Mr. Burke's history class convinced us that women *are* inferior."

But that's exactly what we did talk about. On we went, lampooning the treatment of women in our school. Our classmates caught on that the speech was supposed to be funny and started howling. But the audience stared at us in stony silence. I felt hot and cold and sick, wondering if the adults might boo us off the stage.

The audience responded with tepid applause, but our classmates cheered—a generous response, given that Sharon and I had just thanked them for "being three through ninety" so we could graduate as one and two. As we stepped off the stage, I saw my mother coming toward me. No smile. No congratulations. But neither did she appear upset. "There was something wrong with the microphones," she informed us happily. "None of us could hear a word you said."

That night, at Eric's graduation party, Barry turned to the rest of us and said, "So, do you think Eileen and I should get engaged now, or should we wait until she's finished her first year at Yale?"

No one knew what to say except Eric, who raised his glass and wished us a slurred "mazel tov!" I knew Barry had been drinking. But that didn't alter my perception that the man I loved had just proposed. Did I actually think I would earn my degree, then return home and live with my debate coach? I was simultaneously entertaining both sides of the proposition, the negative and the affirmative. As much as I loved physics, if I couldn't find a man in the wider world who would love me, I would come home and marry Barry.

In August, the Yale catalog arrived in the mail, its cover the exact blue of the summer sky when I took it outside to read it. Studying that catalog was like leafing through the activities at a resort in heaven; you couldn't possibly do everything you wanted to do, but that didn't matter, because you would have so many, many days to do them. Better yet, everyone else in heaven enjoyed these same activities. That's what made it heaven.

By then, even my parents had grasped that I needed to prepare to move out. My mother and I pasted S&H Green Stamps into books, then drove to the redemption center and traded the stamps for a gooseneck lamp, an iron, a wastebasket, and a portable hairdryer with a mushroom-shaped cap that ballooned when you put it on. We drove to an outlet store in Paramus to buy a set of powder-blue plastic valises, each case nesting inside the larger, as if my future were a series of Matryoshka dolls packed one inside the other, to the tiniest female embryo. We stopped at Alexander's and I picked out an orange-and-yellow paisley comforter for my bed. I might have gone to college content with that, but my mother had grown up at a time when a girl needed a wardrobe with which to earn her MRS degree, and she insisted I buy a pair of gray corduroy trousers, an argyle vest that went nicely with the trousers, several turtlenecks, a dress, and a rust-colored polyester cardigan with a belt.

After we got back to Liberty, I piled my purchases in my room and began crossing off the days. Finally, at the end of August, I loaded my hangers, shoes, and linens inside my father's old footlocker, stenciled with his name and rank, as if I were entering the US Army Dental Corps, circa 1943, assigned to fill the nasty teeth of British stevedores in Calcutta, India. (Why did every student in the seventies think she needed such an ungainly piece of luggage to go off to school?) I packed my T-shirts, jeans, corduroy trousers, and turtleneck sweaters inside my new powder-blue plastic valises, which my father arranged in the trunk of our mammoth Pontiac, surrounding them with my new portable Smith Corona, my gooseneck lamp, my iron, my wastebasket, my hair dryer, and my big yellow slide rule, which came in a leather case that made it seem like Excalibur in its scabbard. Then I ran back to my room to make sure I hadn't forgotten anything I wanted to bring to Yale and had left behind every particle of the self I once was and no longer wished to be.

PART II

Surviving Yale

CHAPTER FIVE

Freshman Disorientation

Ours was only the fifth class of women to enter Yale, and the university had decided to house a critical mass of female science majors in Silliman College, as if we were broody hens who might encourage each other to lay golden 100s on our exams. (Students assigned to Timothy Dwight and Silliman live in their residential colleges freshman year; everyone else wallows in communal squalor on Old Campus and takes his or her meals in the cavernous dining hall called The Commons.) Silliman had been founded by a chemist and was the nearest of Yale's residential colleges to Science Hill, but even from Silliman the trek on a slushy, cold New Haven day was daunting enough to make anyone switch to the humanities.

By coincidence, I already knew one of my suite mates. Laurel had attended a parochial school in the Bronx whose team Eric and I debated. I was wary of getting sucked back into forensics, but Laurel's event had been extemporaneous speaking, and when we talked on the phone, she assured me that she wouldn't have time for anything except cramming for her pre-med classes.

The other two members of our suite were Bobbi-Rae, a sweet but graceless pre-med student who came from one of New Haven's working-class suburbs, and Barbara, a petite Manhattanite who spoke in a breathy little-girl lisp and moved with the languid, affected grace of a girl who knows if only she had been born a wee bit taller and hadn't been so much more interested in books than dance, she might have enjoyed a brief professional career as Balanchine's prima ballerina before moving on to study Molecular Biology and Biophysics at one of the better Ivies. I had never met anyone so serene, yet beneath that unruffled exterior she was even more competitive than I was. "I'll get an A in physical chemistry if I have to sleep with Professor Sturtevant!" she exclaimed, Professor Sturtevant being a short, sweet, desiccated chemist who, for dermatological reasons, was undergoing a facial peel and came to class with skin the texture and color of a brick. Guileless, plodding Bobbi-Rae got on Barbara's nerves so badly that, a few weeks into term, I came in to find Barbara yanking open the window and hurling out Bobbi-Rae's stuffed animals as Bobbi-Rae clawed at her from behind. The animals landed in the moat below, and I later saw Bobbi-Rae and her boyfriend scale the brick wall and jump down to rescue the bunnies and bears from a soggy death.

Poor Bobbi-Rae. By the end of that first semester, she had responded to the difficulties of her pre-med classes—and what must have been the insupportable loneliness of being the only working-class kid on campus who had no idea how to hide her working-class origins—by transferring to UConn. By then, I was sorry to see her go, if only because she was the only poor chump below me in our Yertle the Turtle pileup in the swamp.

As for Laurel, she and I barricaded ourselves in our closet-sized room and developed one of those us-against-everyone college friendships that serve as a life raft in stormy seas for all the four years that follow. Not that we didn't have our differences. When I tried to claim the bottom bunk because I walked in my sleep and was afraid I would go for a somnambulant stroll and break my neck, Laurel commiserated, then explained she wasn't remotely athletic enough to hoist herself to the top bunk and was far too myopic to climb down in the morning without her contacts. As she unpacked her electric curlers, straightening iron, bottles and tubes of makeup, and her many pastel sweaters, I felt I had been assigned to share a dorm with Doris Day. When our resident advisor, one of those earthy pioneers who had entered Yale in the first class of women, invited us to her suite and served us carrot cake, Laurel wrinkled her nose and whispered, "She must have used the peels and dirt and thrown away the carrots," while I marveled that anyone in this brave new world would think to bake a cake from carrots.

I had grown up Jewish in the Borscht Belt while Laurel had grown up devoutly Catholic in the Bronx. (That first December, when Hanukkah rolled around, I came back to find birthday candles in a tinfoil holder with a note on which several misspelled versions of the holiday had been crossed out and replaced with a wish that I enjoy a "Happy Handkerchief.") And yet, something told me that being able to bolt the door and huddle with a roommate whose father, before he died, had worked in a factory that printed aluminum cans for Coke, and whose mother was the secretary for a company that manufactured lace curtains (our windows were always far better dressed than anyone else's), would turn out to be more vital to my mental health than latching on to a Roosevelt or a Kennedy. Each of us appreciated how hard the other worked to keep up with her studies—Laurel seemed the only pre-med student on campus who actually wanted to be a doctor. We didn't need to hide how limited our budgets were, although Laurel was under far more financial pressure than I was. I hated that my roommate was tall and blonde while I was short and dark, but being nicknamed Laurel and Hardy was a small insult to bear in exchange for the only friendship I could trust at Yale.

As different as we four were, Laurel, Bobbi-Rae, Barbara, and I were all science majors, as was Erika, who lived next door. In fact, Erika turned out to be a physics major, a discovery I might have found comforting if I hadn't found it so threatening. In the four years Erika and I lived in the same college, so close we often washed at a single sink, I can't recall that we once discussed physics. And as much as I enjoyed the sight of her shining, smiling face, I can't say we were friends. If a person's self-worth derives from being the only woman in a field, how much affection can she feel toward another woman who might challenge that claim to fame? Erika's decision to pursue a bachelor of arts degree rather than the more demanding bachelor of science struck me as cheating. It was as if we had signed up to be marines, and here we were at boot camp, each wearing the same uniform, but Erika got to stay in the barracks and buff her nails while the rest of us jogged fifty miles in the rain.

The first night of orientation, the dining halls weren't open, so Erika suggested we go out to eat Chinese. I allowed my hall mates to do the ordering, then tried not to show my misery when I bit into one of the red, dagger-shaped objects in the dish they were all enjoying. But my distress was impossible to conceal when I saw the bill—my share came to half my month's allowance—or my amazement when the other girls pulled out their credit cards. Credit cards? All I had was a letter from my father's bank in Liberty assuring whomever was in charge of the bank in New Haven at which I would be opening a new account that I was "the daughter of a well-known and respected family in our community" and asking "whomever it might concern" to extend whatever courtesy was in his power to provide. (When I showed this document to a teller in New Haven, he stroked his chin and asked if he might share it with his associates.)

Even the nice girls baffled me. One woman—she lived on the other side of the fire door—was innocent and kind, with corkscrew red hair, wide, startled eyes, a drawling Southern accent, and an open, freckled face. Elva invited me to join her in a game of squash, but the only squash I had ever heard of was zucchini, and the sight of the windowless underground box in which the game was played gave me the sweats. Elva mentioned she was trying out for the women's basketball team and invited me to come. Excited by the prospect of proving myself a star, I met Elva at the gym. But playing a version of basketball in which we girls had been allowed to use only half the court hadn't prepared me for running drills the length of the gym while the coach screamed and blew her whistle. Ten minutes into tryouts, I needed to dash into the locker room to avoid throwing up on the court.

Having spent much of my childhood trying to prove I was as fast as any boy, could hit a ball farther and smack a forehand harder, here I was surrounded by female athletes, and I couldn't last ten minutes without puking up my lunch. Who were these Amazons? One of Elva's suite mates woke before dawn, ran six miles to the river, rowed crew, then dashed back to Silliman to shower and change before class. (This student and her teammates protested the lack of a women's locker room at the boathouse by marching into the athletic director's office and pulling off their sweatshirts to reveal TITLE IX markered across their chests.)

After we got back from the Chinese restaurant, the boys in the next entryway invited us to a party. I put on my new gray corduroy trousers and matching gray wool vest, then ventured next door, accepted a beer, and sat on a bed and sipped it. One of our hosts flopped on the bed beside me. Desperate to make conversation, I asked where he had gone to high school.

"Phillips Exeter," he informed me flatly.

Oh, I said, and where was that?

"You've never heard of Phillips Exeter?" If we had been cartoon characters, his eyeballs would have popped out on springs. "It's only the most famous private academy in the world."

But the only private academy I had ever heard of was Valley Forge, where my parents had threatened to send my brother. "You went to a military academy?" I said, at which my host snorted in disgust and left.

Years later, when I taught at Harvard, I met an African American student who had grown up in Harlem and won a fellowship to Phillips Exeter. Just before he was due to leave home, Tarik and his mother received a list of items he needed to bring. One article—a navy-blue jacket—puzzled them both. The only jackets anyone in Tarik's neighborhood wore were blue windbreakers, and the only kids who wore them were gang members. Then again, if all the students at Phillips Exeter were supposed to wear blue windbreakers to dinner, who was Tarik to argue? At least windbreakers weren't expensive. Which is how he came to show up for his first meal wearing a blue windbreaker, only to discover that every other boy was attired in khakis, Top-Siders, and navy-blue blazers with gold buttons on the sleeves.

The dislocation I experienced moving from the Borscht Belt to Yale wasn't nearly as severe as the dislocation Tarik must have suffered in moving from Harlem to Phillips Exeter. But I could understand how an article of clothing as innocent as a windbreaker might cloak its wearer in humiliation. My first day on campus, I proudly bought a lined, waterproof blue jacket with YALE stamped across the chest in white. Seeing my purchase, one of my hall mates informed me that clothing emblazoned with the university logo was meant for tourists. People who went to Yale were too modest to wear articles of clothing that might shame people who weren't fortunate enough to attend the school. Mortified, I considered what to do. I didn't own another warm, waterproof jacket, and I had just spent my last fifteen dollars buying this one. Flustered, I took a pen and scribbled over the white lettering, only to realize the only thing more gauche than wearing a jacket with YALE emblazoned across the chest was wearing a jacket with YALE bizarrely scribbled out with ballpoint pen.

Here's how little I knew about the dress code. When a friend explained that one of our classmates had amassed the largest and most impressive collection of Americana in the world (I needed to ask what *Americana* meant) and had bought a defunct bank in Texas so he could keep his collection in the vault, I asked why, if our classmate was so rich, he walked around in ratty jeans and a parka mended with duct tape. Well, my friend said, people who came from "old money" signified their disdain for status by driving beat-up Volvos and walking around in parkas mended with duct tape, at which point I understood that even if I took a razor to the expensive new parka I had persuaded my parents to buy and then patched those rips with duct tape, I would never fit in at Yale, just as if I took a razor to my skin, the blood that flowed would not be blue.

There were plenty of Jews at Yale. They just didn't look, talk, or act like any Jews I had ever met. If my gentile classmates were ignorant of the Catskills, my Jewish classmates made a point of telling me that their parents wouldn't have been caught dead vacationing at a dump like Grossinger's, which, compared to my family's hotel, was Buckingham Palace. After living in a town where one of three people was a full-bearded, black-caftaned Hasid, I spent four years without seeing a single yarmulke. There were kids I thought might be Jewish, but they didn't have Jewish names, and kids who had Jewish names but weren't Jews—a mystery I solved when a classmate admitted his father had converted to Christianity so he could be promoted at a firm that didn't hire Jews, and another classmate revealed his father had changed his name to effect the same result. I was shocked my Jewish classmates had attended schools with names like Trinity or Friends. They spoke in quieter, softer voices than I did, and their hair was miraculously soft and straight. The one or two kids who struck me as jewy Jews, like me (their voices—ugh! their hair!) I avoided as best I could.

Being a Jew at Yale didn't hold a person back. But being a jewy Jew at Yale made me even more self-conscious than I otherwise would have been. For the first time, I heard the honk and squawk of my Borscht Belt howl; a lifelong show-off, I became reluctant to open my mouth. I would use a Yiddish word, only to watch it fall at a classmate's feet like a deflated beach ball. Ashamed of my curly hair, I took my sister's advice and rolled it on juice cans, then sat under my dryer for hours; if you look at my photo in the freshman face book, I have the long, slinky locks that Monica wears on *Friends*.

Like most kids, I wanted to become a new person when I got to college. But that newness didn't include purging myself of my Jewness. I was searching for something I would now call grace, not understanding that grace, by definition, cannot be hunted. At orientation, we filed into Woolsey Hall to hear Kingman Brewster's welcoming address. As I sat in an auditorium whose name brought to mind some fat cardinal notorious for beheading queens, I looked up at the university's renowned three-story Newberry Memorial Organ, listened to my fellow freshmen joke about how Yalies had more impressive organs than their counterparts at Harvard, and thought about how intimidated my parents would have been to sit in these same pews, cowed by an instrument they associated with castrated boys inquiring in musical Latin, "Did the Jews kill Christ?," to which the answer was a resounding *yes*!

"This is primarily a place for learning," Brewster told our class. "But not all learning is in books or laboratories or classrooms. You probably have not been as free before. You may not be as free again. Enjoy the privilege of doubt. Make the most of it."

The speech became famous among Yalies later. But having sacrificed so much to read those books and attend those lectures and carry out experiments in those labs, I wasn't prepared to hear I should pursue some nebulous idea of freedom rather than buckle down and study physics.

My schedule that first semester—advanced physical chemistry and its lab, advanced calculus, the first term of the physics sequence for majors, introductory economics, and Western lit—was the equivalent of a novice weight lifter attempting to bench press a refrigerator above her head. The one concession I made to the academically deficient diet on which I had been raised was that I didn't sign up for the physics sequence intended for students who already had discovered a flaw in the theory of relativity.

The only other physics majors I knew were the ones in Silliman. One Sillimander, a Minneapolis boy named Jim, was brave enough to sign up for intensive physics, along with Mikis, who had straggled in weeks late, haggard and unshaved, having survived a war I hadn't known was going on in a country I had never heard of. Not willing to risk academic suicide by joining them, I consoled myself that I wasn't taking Physics for Poets, as if poets were the lowest form of life.

Visiting the co-op to buy books for that first term's courses, I was as astonished as a child who has been raised on Mars bars and M&M's and then finds herself in a market that sells chicken, beef, fish, eggs, watermelons, cherries, and asparagus. I had been inside a bookstore only a few times in my life, and here I was loading my arms with *The Iliad* and *The Odyssey*, *Oedipus Rex*, and *Faust*. Best of all: here was the textbook I would be using for Physics 14A. To say my edition of Halliday and Resnick's *Physics* was big isn't to convey how big a book it really was. It was, quite literally, the heaviest, thickest book I had ever seen, not to mention that the word *PHYSICS* appeared in giant orange letters across the cover.

My pleasure in carrying my big new *PHYSICS* textbook to the register was offset by the shock of seeing the clerk ring up the price.

Twenty dollars? With the other books in my load, that brought the total to two hundred dollars (which happens to be the price of the tenth edition of that same physics book today). Still, I was grateful to be living in a place where a student could carry a book that thick without being afraid another student, to amuse his friends, would demonstrate the laws of Newtonian mechanics by grabbing it and dropping it on her head. By the time I had lugged my purchases to my room, my arms were ten feet long. I stacked my new textbooks on my desk, then opened them one by one, saving my physics book for last.

In high school, my physics book had been composed of words, while the same chapters in my college textbook were filled with diagrams and equations. Rather than see this as a warning—not only had I not learned the material in this book, I had not learned the material I would need to learn the material—I couldn't wait to confront the first real academic challenge I had ever faced.

And now? How do I feel opening my old copy of Halliday and Resnick today? At first, I am as frightened by its massive size and cold, unyielding formulas as any poor numskull who didn't spend four years of her life earning a BS in physics. I see penciled on the inside cover the binomial expansion of $(x + y)^n$ and am as mystified as if I had found a boy's name embellished with hearts and arrows and couldn't remember the face that fit that name.

The book is even more sexist than I recall. The scientists in the photos are all lab-coated males. To illustrate the weight of a passenger in an elevator accelerating up or down, the authors show a cartoon version of Albert Einstein holding an anvil at crotch level, its imposing arrow-headed vector dangling below his waist. The problems at each chapter's end involve bats and balls and bombs. Even the disembodied hand that illustrates the right-handed rule by which to find a vector's product is hairy and thick.

The sight of one particular diagram triggers a flashback to an entire week I spent plotting the path of a ball tossed not from level ground but from halfway up a hill. Pad after pad of legal paper disappeared as I sat beneath my poster of *Christina's World* (which I was sure portrayed a passionate young woman looking up the hill she was about to climb, not knowing the model for Christina was a woman so crippled, the only way she could have reached the top would have been to crawl there), struggling to calculate gravity's effects on the ball's trajectory. The night before the problem set was due, I stayed up late, then gave up in despair and went to sleep, only to find myself slogging through a sludge of sines and cosines in my dreams. Toward dawn, I decided I had found the solution; rolling out of bed to write it down, I fell from the top bunk and landed on my roommate's desk. Laurel bolted upright and hit her head. The solution I had dreamed was meaningless, and, for the first time in my life, I was forced to hand in my homework blank.

But all that came later. The first morning of that first semester, I could barely keep down my breakfast. Could anything have been more exciting than carrying a pristine notebook embossed with *Lux et veritas* to a lecture hall where I would finally begin the life I had been waiting eighteen years to start? My status as one of only two women in the auditorium struck me as less frightening than erotic: it was like going to a movie with 118 dates. I was even more excited when the professor turned out to be the same dark, bearded young man whose class I had visited the spring before.

That excitement turned to alarm as he raced across the stage, weaving equations I couldn't unravel and telling jokes the humor of which eluded me. I hadn't understood anything he had said the spring before, but I figured I had arrived at the movie late, and if I came in at the beginning, I would understand everything I had missed. The truth is, if you don't know the language in which a movie is being shown, you won't have any better grip on the plot if you come in at the beginning than at the end.

The boy to my left leaned back and stopped taking notes. "Jesus," he sighed. "We covered this shit in high school."

Studying physics, as it turned out, didn't involve the contemplation of space and time, as I had hoped, but problems such as: "A rifle with a muzzle velocity of 1500 ft/sec shoots a bullet at a small target 150 ft away. How high above the target must the gun be aimed so that the bullet will hit the target?"; and, "A dive bomber, diving at an angle of 53° with the vertical, releases a bomb at an altitude of 2400 ft. The bomb hits the ground 5.0 sec after being released. (a) What is the speed of the bomber? (b) How far did the bomb travel horizontally during its flight? (c) What were the horizontal and vertical components of its velocity just before striking the ground?" The equations weren't complicated. But there was something I wasn't getting, some deeper intuition about the way swings swing and pulleys pull and centripetal (or maybe it was centrifugal) forces act on any ant misguided enough to be sitting on the rim of a vinyl LP spinning at 33 1/3 rpm. I felt like someone who has been ordered to fix a car without having the least idea how an engine works.

Nor did I have experience with homework being hard. In high school, I had ripped through the assignments for one class while the teacher in another class jabbered at the board. Now, I sat in the cramped room I shared with Laurel, randomly plugging data into this or that equation, hoping the result would match the answer at the back of the book. I got so frustrated I went up to Mikis in the dining hall and asked if he could help. After weeks of access to the all-you-can-eat serving lines, he was less gaunt than when he had straggled in from whatever war zone he had escaped. He had shaved his beard and cut his hair. But he still seemed older than the rest of us. For a fresh-faced American girl to claim she was majoring in the same subject as he was must have struck him as an insult. He took a drag on his cigarette, then confirmed I had no intuitive feel for science. What I needed to do was to think about the *physics* of the problem. To demonstrate, he leaned back, closed his eyes, and let smoke billow from his nose.

As angry as I was, I knew he was right. But where could I obtain this talent?

Mikis shook his head. If you weren't blessed with physical intuition, there was no store in which you could buy it. Still, all wasn't lost. "Don't worry, Eileen," he said. "When I am big professor, you can be my assistant and run my lab."

Of course, more than a few of the 118 boys in that introductory physics class had trouble keeping up. But I didn't know that then, any more than I knew most of the boys worked on their problem sets together; in assigning all the female science majors to Silliman, the administration had prevented us from stumbling on the boys doing their problem sets together in the all-male entryways on Old Campus. Then again, I wouldn't have had the courage to ask those boys for help. Why would I have let them know how desperately behind I was? As to seeking out my professor, why would I expose my ignorance to such a brilliant man?

That left only Erika, but her problem sets came back with even lower scores than mine. How could she remain so insanely confident? No matter how difficult a class, Erika was never ruffled. She was like some small, pugnacious dog—a female version of the Yale mascot, Handsome Dan—who finds herself in the ring with a Doberman or Rottweiler and dances about, snapping and strutting so brazenly that the larger, fiercer dog becomes suspicious and backs away.

I spent the rest of that semester holed up in the basement of Kline Science Library searching through book after book in the hope of finding a sample problem similar to the one I needed to solve for that week's homework. Or I sat in my room, reading the same difficult passage in Halliday and Resnick over and over. If not for Laurel's presence at the adjoining desk, I might have lost the power to communicate with another human being. Laurel was struggling with her pre-med courses, and we found solace in each other's sighs, the squeak of a highlighter against a page, the crackle of a sheet of graph paper being crumpled and tossed in my much-used wastebasket. Laurel was more accustomed than I was to finding comfort in her female friends, and every now and then she would snap her biology textbook shut and say, "Okay, roomie, that's it for me," and we would brew Lipton Cup-a-Soup in our hot pots and talk about her boyfriend, Bill, and how he was doing at MIT, or why Barry hadn't responded to my letters. Then we went back to studying, only to take another break to walk to Wawa's, where we would buy a gallon jug of grapefruit salad so tart my lips pucker at the memory, or descend to The Buttery in the basement, where we ordered coffee and English muffins to revive our brains.

I didn't meet another physics student who was suffering as much as I was until we took our midterms. As the class crowded around

Professor Zeller's door, I stood on my toes and saw a big red 32 markered beside my name. Thirty-two? Did numbers go that low? Wobbling away, I came upon my classmate Ronaldo, who seemed equally as distraught as I was. Ronaldo was yet another Yalie whose ancestry puzzled me: he had grown up in Mexico, but he had a fair complexion and spoke English without an accent. Unlike most Yalies, there was something hangdog and vulnerable about Ronaldo; maybe it was that his mustache drooped and he walked with a romantic limp (I never asked why—maybe he'd had polio as a child). When it turned out his score was barely higher than mine, I felt oddly euphoric; for the first time, I could join a boy in complaining about a grade.

Back at my suite, I listened to Ronaldo agonize about how to tell his father he might not pass physics, which, as an architecture major, he needed for the upper-level requirements. That anyone's parents might want him to succeed in physics struck me as bizarre. My father rarely wrote letters, but I still have the note he sent in response to my anguished revelation that I had gotten a 32 on my physics midterm. "It was nice speaking to you the other nite," he starts, then immediately advises that I have "picked too tough a schedule" and bets that "a couple of more grades like the one in Physics will turn you from that course."

If you had asked my parents, they would have said they wanted me to succeed. But they didn't want me to set my sights too high, only to be disappointed. They didn't want me to attract the malevolent gaze of the Evil Eye. Most of all, they wanted me to be able to earn a living until I married a man who would earn my living for me, and physics seemed unlikely to accomplish either goal. There *were* no women physicists, and even if there had been, no man in his right mind would have married one.

Well, if everyone expected me to major in English, why not go ahead and please them? Deciding to drop physics seemed as easy and alluring as falling asleep in snow. I trudged up Science Hill to ask my professor to sign my withdrawal slip. To get to Gibbs Physics, one needed to battle one's way past the Kline Biology Tower, which resembled a giant Tootsie Roll on its end. The pillared arcade surrounding Kline funneled the wind, so I was forced to bend ninety degrees to cross it, then summon all my strength to open the metal door to Gibbs. Inside, I took the elevator to Professor Zeller's floor, then navigated corridors lined with thick conduits for wires, humming generators, photos of the all-male faculty, and corkboards with the answers to problem sets for upper-level courses (if those were the answers, I didn't want to see the questions). I found my professor's office and convinced myself to knock because returning without his signature would consign me to even worse humiliation.

"Yes?" he said. "Come in."

I was so flustered I could barely talk, but he smiled encouragingly, and I managed to stammer that I had gotten a 32 on the midterm and needed him to sign my drop slip.

"Why?" he asked, as if the answer weren't obvious. When I didn't say anything, he told me that he had gotten Ds in his first two physics courses. Not on the midterms, in the *courses*. The story sounded like something a nice professor would invent to make his least talented student feel less dumb. In his case, the Ds clearly were aberrations. In my case, the 32 signified I wasn't any good at physics.

That I didn't drop the course can be attributed to what my professor told me next. "Don't pay attention to how anyone else is doing. Just swim in your own lane." Seeing my confusion, he said he had been on the swimming team at Stanford. His stroke was as good as anyone's. But he kept coming in second. "Zeller," the coach said, "your problem is you keep looking around to see how the other guys are doing. Keep your eyes on your own lane, swim your fastest, and you'll win."

I gathered this meant he wouldn't be signing my drop slip?

"You can do it," he said. "Stick it out."

For once, being female was an advantage. I developed such an overpowering crush on Professor Zeller that I stayed in the course for him. As he raced around writing equations on the board, I lost much of what he said because I kept fantasizing about what those swimmer's shoulders would look like without a shirt. But my reluctance to disappoint the object of my crush prevented me from giving up. Week after week, I struggled to do my problem sets, until, by the end of term, they no longer seemed incomprehensible. Which is why, the deeper I tunnel now into my copy of Halliday and Resnick, the more equations I find festooned with comet-like exclamation points and theorems whose beauty I noted with exploding novae of hot-pink asterisks. The markings in the book return me to a time when, sitting in my cramped room, I suddenly grasped some principle that governs the way objects interact, whether here on Earth or light years distant, so my heart raced and my skin grew hot and my mind expanded to contain the universe, and I marveled that such vastness and complexity could be reducible to the equation I highlighted in my book. Could anything have been more thrilling than comprehending an entirely new way of seeing, a mystical layer beneath the real, a reality more real than the real itself?

My new powers of understanding might have flowed from nothing more than Professor Zeller's voice murmuring seductively in my head: *You can do it. Stick it out.* Or those new powers might have been attributable to the final piece of advice he delivered before I left his office, still clutching my unsigned drop-slip: *The best way to understand the physics behind a problem is to sit down and study Feynman.*

Most nonphysicists were introduced to Richard Feynman's unique brand of intuitive, gut-level science during the 1986 hearings to investigate the explosion of the space shuttle *Challenger*. After days of debate as to whether cold weather might have caused the shuttle's rubber O-rings to lose resiliency, Feynman ended the argument by dropping an O-ring in a glass of ice water and telling the commission, "I took this stuff that I got out of your seal and I put it in ice water, and I discovered that when you put some pressure on it for a while and then undo it, it doesn't stretch back. . . . I believe that has some significance for our problem."

But even when I was a student, Feynman was larger than life—the prankster who spooked his fellow scientists at Los Alamos by guessing the combination of the safe that held the secrets to the atomic bomb; the last living link to Einstein, who attended Feynman's first lecture while Feynman was a student at Princeton; the winner of a Nobel Prize for his theory of quantum electrodynamics; and the inventor of a stunningly simple set of diagrams that allow physicists to keep track of the complex interactions among particles and antiparticles in space-time. Rangy, tousle-headed, iconoclastic, Feynman exuded what passed for cool among physicists of the era—he played the bongos, was an unabashed devotee of topless bars, and had the perfect excuse for his womanizing in the tragic death of his teenage sweetheart, who had succumbed to TB while her young husband was saving the free world at Los Alamos.

When my professor advised that I "sit down and study Feynman," he was referring to the three red volumes that contain the introductory lectures Feynman delivered at Caltech in the sixties. The first page of each book displays a photo of the Great Man Himself, so every time I sat down to read *The Feynman Lectures on Physics*, I was reminded of what a real physicist should look like, his face hawklike in intelligence, everything about his demeanor vibrant and energetic, hands moving so fast you could almost see the sound waves rising from the drums. A female student sitting down to read the *Lectures* had a hard time putting herself in the shoes of the brash, bongo-playing womanizer who had written them.[1] But that was what the lectures were about. Feynman's goal wasn't to teach you this or that physical law. Only a fool would attempt to memorize all the laws of physics—even if you succeeded, you would be left with a head full of laws rather than the intuitive understanding of nature that would allow you to derive the equations you had forgotten. Or better yet, develop new ones. What the author of the *Lectures* wanted to teach you was how to be Richard Feynman. The difference between reading an ordinary physics book and reading Feynman was the difference between taking lessons at an Arthur Murray studio and hanging around while Rudolf Nureyev choreographed a new ballet.

---

1. When the first female physics professor that Caltech hired—and then neglected to promote—sued to be granted tenure, Feynman took her side. But like most scientists of his era, he didn't think twice about using sexist language. When he won his Nobel, he told the audience that falling in love with his theory of quantum electrodynamics had been like falling in love with a woman, "[which] is only possible if you do not know much about her, so you cannot see her faults." By now, he said, the theory had become "an old lady that has very little attractive left in her and the young today will not have their hearts pound when they look at her anymore. But, we can say the best we can for any old woman, that she has been a good mother and she has given birth to some very good children."

The comparison to Nureyev isn't far-fetched. According to his biographer, James Gleick, colleagues who watched Feynman concentrate on a problem "came away with a strong, even disturbing sense of the physicality of the process, as though his brain did not stop with the gray matter but extended through every muscle in his body." Once, when Feynman was an undergrad at Cornell, a classmate came in and saw him rolling around on the floor, which was Feynman's way of doing his homework. For Feynman, the elements of nature "interacted with palpable, variegated, fluttering rhythms."

This was what Mikis had tried to tell me. When confronted with a problem, you shouldn't just plug the data into some equation. Rather, you should close your eyes and visualize the objects in the problem moving and interacting. What usually came first for Feynman was the image, the dance. Only after he had gotten the picture clear in his head did he attempt to communicate his intuition via math.

The complete set of Feynman's lectures is so expensive that even today, I can't afford to buy one; I need to visit the undergraduate library at the University of Michigan, where I teach creative writing. I haven't ventured into the science section of a library in decades, and the titles there seem forbidding. I can barely remember the exhilaration I used to feel burrowing through the stacks where so few Yalies dared to go, the satisfaction I took in knowing I was privy to secrets so few mortals knew. And there they are, three sets of the three red tomes, with empty spaces where copies have been checked out by undergraduates as eager today to transform themselves into clones of Richard Feynman as I was in the seventies. I stack the *Lectures* before me, and, with the reverence of a lapsed Jew unrolling a Torah, I open volume 1.

Immediately I am reminded why I once worshipped the god I did:

If, in some cataclysm, all of scientific knowledge were to be destroyed, and only one sentence passed on to the next generations of creatures, what statement would contain the most information in the fewest words? I believe it is the *atomic hypothesis* . . . that *all things are made of atoms—little particles that move around in perpetual motion, attracting each other when they are a little distance apart, but repelling upon being squeezed into one another.*

In reading that one sentence, a student has gone from being an ordinary ignoramus to the sole human being who, if civilization were destroyed, could re-create the most magnificent achievement of the race. I consider working through all three volumes, but the eighteen months I would need to accomplish that goal would constitute one-twentieth of the span allotted me before my atoms obey the laws of entropy covered in chapter 40.

Instead, I content myself with reading a slim volume called *Feynman's Tips on Physics*, published in 2006 as a supplement to the trilogy. As much as I used to beat myself up for not understanding every paragraph of the *Lectures*, it turns out very few freshmen in Feynman's original (all-male) class at Caltech were able to follow everything their professor said. A fifth of the students stopped coming. (As word of the lectures got around, doctoral candidates and professors began taking the empty chairs.) To help the slower students, Feynman held a series of review classes; many years later, someone located a recording of those sessions and published the transcript. Sitting in Michigan reading *Feynman's Tips*, I find myself tearing up; it's as if I had set off to find the Seven Lost Cities of Gold and was forced to abandon my quest because I couldn't find the waterhole from which everyone else seemed to drink, but here I am holding the map that indicates the location of that oasis with a giant X.

For all I imagined Feynman to be a bongo-playing jerk, he seems to have understood how debilitating it can be to graduate as valedictorian of your high school and then score a 32 on your first physics midterm. Addressing a roomful of students despondent because they found themselves at the bottom of their class, Feynman reminds them that even at the most rigorously competitive university, half the class must, by necessity, earn below-average grades. But he isn't finished. Discovering you are in the bottom half of a physics class at a top-ranked university begets an *emotional* problem. There's no *logical* reason to be discouraged. "Therefore," he says, "I am making this review purposely for the people who are lost, so that they have still a chance to stay here a little longer to find out whether or not they can take it, okay?"

Having quelled his students' fears that they got admitted to Caltech by accident, Feynman attempts to convey what it means to use physical intuition to solve a problem. By leading them through examples of how he perceives the physical world, he hopes his students might yet catch on. The examples Feynman offers are so effective that even if you flunked general science in high school, you can appreciate what he's doing. (If you flunked general science and ended up working as a carpenter, you will have exactly the sort of physical intuition Feynman is trying to impart.) Imagine you are faced with a concrete block supported on top of two steel rods that are arranged in a giant inverted *V*. Each rod has a roller at the bottom, so the block moves up or down as the rods roll closer or farther apart. Now think about the ways the block distributes its weight through the rods to the rollers. If you shove the rollers closer to each other, so the block is way up high, a lot of the force will be exerted downward on the rods, with very little force exerted sideways.

"If you can't *see* it," Feynman says, "it's hard to explain *why*—but if you try to hold something up with a ladder, say, and you get the ladder directly *under* the thing, it's easy to keep the ladder from sliding out. But if the ladder is leaning way waaaaay out, so that the far end of the ladder is only a very tiny distance from the ground, you'll find a nearly infinite horizontal force is required to hold the thing up at a very slight angle.

"Now, all these things you can *feel*. You don't *have* to feel them; you *can* work them out by making diagrams and calculations, but as problems get more and more difficult, and as you try to understand nature in more and more complicated situations, the more you can guess at, feel, and understand *without actually calculating*, the *much* better off you are!"

The funny thing is that even though I had never worked construction and would have been doubly lacking in confidence as the only woman in that review class, I did have some of the intuition Feynman was attempting to impart. I would have had a lot more of it if I had grown up building tree houses and tinkering with car engines, or if I had been allowed to take shop class instead of home ec. But even I could have told you where you needed to push to keep that ladder from collapsing. What I didn't have was the courage to trust my intuition.

Take the principle of least action. Most physical systems settle into whatever state requires the least energy to maintain. That's how a chain figures out which curve to follow when you suspend it between two poles, or a soap film decides what shape bubble to form when you lift a wand from a bowl of suds. A similar principle governs the path a light ray travels when it moves from one medium to the next—say, when you are looking at a fish. Rather than take a straight line from your eye to the trout, the light enters the water at an angle. You could say the light is applying the same strategy a lifeguard uses when she sees a swimmer drowning a long way down the beach. The lifeguard knows she can run faster than she can swim, so if she takes the most direct line from her chair to the person flailing in the waves, she will spend only a short time running on the sand and far too long slogging through the water. But if she runs to the point directly opposite the drowning man, then jumps in and swims perpendicular to the beach, she will add so much distance to her race along the sand that she will still waste too much time. The fastest strategy is to chart a path somewhere between the one that would take you directly to the water and the path that would have you running along the beach until you are opposite the swimmer. To find that optimum path, a lifeguard would need to sit down and use calculus; instead, she relies on training and intuition to save the swimmer.

But how does a beam of light calculate where to enter the water so it can reach the fish and travel back to your eye in the shortest time? How does the soap bubble know which shape will minimize the energy required to maintain it? Even as a freshman, I suspected that if I could answer these questions, I would comprehend something fundamental about the universe. As I later learned, a similar set of questions bugged Feynman; pursuing the answers led him to the discoveries for which he won the Nobel Prize. Being baffled by the same questions that baffled a Nobel Prize winner doesn't make a student a genius. But it's a sign she is thinking creatively. Why didn't I pursue this line of investigation? Whenever I thought of a question, I assumed the answer must be obvious, or someone had already solved it, and if someone hadn't solved it, who was I to think I could? It never occurred to me to visit a professor and ask how light rays know which path to follow. How would I have phrased my question? "Professor, how is a ray of light able to sniff out the shortest path?" To have used such language—which is pretty close to the language Feynman used—would have opened me to the charge of relying on some kind of touchy-feely, womanish anthropomorphization of a light ray.

In his review lectures, Feynman assured his students they didn't need to understand every topic he covered in the course. Most people aren't savvy enough to figure out what's interesting to them and pay attention only to that, he said. As it turned out, I had an intuitive feel for the most abstract questions of theoretical physics. The subjects I liked—quantum mechanics and gravitation—I happened to be good at. So why did I think I needed to be good at ballistics and mechanics? I should have taken a cue from Mikis and shrugged off my incompetence by saying that if I needed to build some contraption to test a theory, I would hire a male technician to build it for me.

Even if I didn't have the benefit of Richard Feynman's tips, I did have Michael Zeller and his advice that I ignore the swimmers in the other lanes. By the end of that first semester, the boys who had attended prep school no longer could rely on their advantages. Endurance became more important than a sprinter's speed. I studied for the final, and when I looked at the list on my professor's door, I saw that I had scored the third-highest grade in the class, which meant that even with my 32, I earned a B. When I stopped by Professor Zeller's office to thank him, he smiled his Paul Newman smile and his blue eyes beamed. "Thatta girl!" he said. "See you next semester!"

And the next semester, I earned an A.

What stuns me isn't that I got off to a shaky start and, much to my credit, studied hard and pulled ahead. It's that the version of the story I always tell is how I was so terrible at physics that I earned a 32 on my first exam and never caught up. For my entire four years at Yale, I saw myself as *handicapped* or *behind*.

The truth is, the less a subject had to do with the visible world, the more talented I was at solving problems. I was nearly as far behind in calculus as I was in physics. But I wasn't the only woman in the class, so I felt more comfortable asking questions. And I discovered I was far more gifted at imagining the behavior of an infinite number of infinitesimally tiny quantities than of bullets, bats, and cannons. The same held true for chemistry. I had an easier time picturing the dance of a nucleus and its electrons than the path one billiard ball might follow after hitting another. Laurel was in the class, along with Barbara, Bobbi-Rae, and a frail, shy psychology major named Rochelle, who struck me as the kindest, least competitive woman I knew at Yale. Our textbook showed no chemists of either sex, and the equations described a universe in which the greatest division lay not between female and male but positive and negative charges.

The disaster came in chemistry lab. After so many years of carrying out my own lame science projects at home, I was allowed to perform experiments in a laboratory more substantial than one that could be folded up in a case. Maybe if that first class hadn't fallen on Rosh Hashanah, or I hadn't worn a dress and pantyhose, or I hadn't felt so humiliated standing there with my mottled, smoking legs being stared at by sixteen male classmates, I might not have come to see myself as unable to pick up an instrument or a piece of glassware without causing some catastrophe.

As it was, I accepted the role of the perky but bumbling co-ed and played it to the hilt. Even though I cried, genuinely distraught, each time I caused a mishap, I related each incident with relish to my friends back home. Everything about my childhood had prepared me to play this part. Hadn't I spent junior high and high school practicing how not to look smart in front of boys? Hadn't I grown up giggling at Gracie Allen and loving Lucille Ball? Freshman year, I went to see *Bringing Up Baby* with my fellow physics major Jim. He found the movie to be annoying, but I took home the message that bringing down a T. rex skeleton was a highly effective way for Katharine Hepburn to persuade Cary Grant to marry her.

As much as I worried that the guys in my lab saw me as a clumsy broad, I also got the sense they liked me. My partner, a stocky rugby player named John, thought I was so adorable that one day, in the middle of an experiment, he threw me across his shoulder and carried me down the stairs. Larry, the lab assistant, seemed dismayed by my many screwups, but he paid me more attention than he paid my male classmates. If nothing else, my disasters made the boys feel better about their own fumbling attempts to get their experiments to succeed.

What none of them guessed was that I was the one bringing up the curve. When the time came to write up my lab reports, I provided more extensive error analyses than the handouts required and offered more comprehensive explanations of the theories behind the experiments. None of this came easy. My parents balked at spending two hundred dollars on one of the amazing new Texas Instruments calculators most of my classmates owned, so I carried out my calculations on my slide rule, spending an extra twenty hours a week writing up my error analyses. But even with my clumsiness at the bench, I earned all As.

That was the double bind that strangled me. If I did poorly, I would prove women never did finish their degrees in science or math; if I succeeded, I would be even more unpopular than before. Bad enough to be a girl who had gotten all As in high school; how much more of an oddball would I be if I earned all As as a physics major at Yale? The only way to escape this paradox was to do well on my exams and lab reports but remain quiet in class and present myself as a lovable if clumsy clown in lab.

My best performance came at the end of the second term of P-Chem. My three lab mates and I had just spent a month synthesizing a compound whose name sounded like *cis-boom-bah*. I was to return to the lab that night and pour the liquid our group had so painstakingly synthesized through a suction filter to distill what should have been 0.30 grams of a purple crystal. Stirring the sediment with a glass rod, I scratched the filter and . . . whoosh! . . . the hose sucked a month's worth of work into the waste receptacle. Terrified that my partners would no longer be amused by my ineptitude, I poured the contents of the waste drum through a new filter. The next day, as Larry and my teammates watched in amazement, I caused 20 grams of fluorescent orange crystals to materialize in the funnel.

The script was writing itself. Wasn't this the plot of every screwball comedy I had ever seen? Perky but absentminded co-ed makes a shambles of the lab. Then she accidentally discovers a strange new compound and works all night with her lab instructor to solve the mystery of the compound's essence. Over coffee at dawn, they fall in love. "Whose name goes first on the article?" the co-ed asks. "We'll just have to publish under the same last name," the instructor replies, laughing.

Instead, Larry ordered me to make up for my blunder by figuring out the structure of the compound I had synthesized. He didn't have time to guide me through the analysis but offered to let me use the lab after hours. I had visions of figuring out that the orange compound was a substance never before seen on Earth, with properties that cured humankind's every ill. Then I realized I couldn't have identified a teaspoon of Kool-Aid unless I had seen the instructor ladle it from the jar. The prospect of figuring out which tests to run and how to run them made me ill. So I didn't pick up that lab key. I was afraid I would fail the course. But when the grades were mailed home, I saw that the instructor had docked me only a few points for my clumsiness, and I still received an A.

The trouble was, I *felt* like a failure. Whenever I heard the word *chemistry*, I saw the contents of that filter getting sucked into the waste receptacle, saw those bright orange crystals of cis-boom-bah appearing in the funnel when my teammates were expecting purple. I felt ashamed I hadn't shown up to identify the mysterious orange compound, even though a much more complex set of skills would have been required than I was likely to possess. From then on, whenever I stepped into a lab, I thought of myself as a danger to myself and others.

CHAPTER SIX

Too Much Male Hormone

As much as I pined for Barry, his failure to follow up on his drunken marriage proposal at my high school graduation persuaded me that I ought to find someone my own age to love. With the ratio of women to men at Yale one to three, you would have thought that wouldn't be hard. But I spent most of my waking hours in the science library, or in classes with other physics majors, none of whom asked me out, either because they thought I was dumber than they were, or because they suspected I was smarter.

Other than that, the only boys I knew were the ones who lived in Silliman, and my hopes of dating them vanished the night I heard eleven of my dinner companions discuss the road trips they were planning that weekend to Smith, Wellesley, UConn, and Albertus Magnus, which the boys smarmily referred to as Albertus Mattress. With women at Yale so scarce, I supposed my male classmates couldn't be blamed for seeking companionship elsewhere. But I wasn't that bad looking. The only reason I could see that I wasn't datable was that I was majoring in a subject they saw as threatening. As they went on discussing their upcoming road trips, I needed to bite my tongue to keep from asking, "What am I, chopped liver?" Half the table wouldn't have known what chopped liver was, and the other half would have pretended they didn't.

. . .

One of the other reasons I didn't seem to be having fun was that my allowance was too small to allow me to keep up with my more generously financed classmates. Tired of counting every dime, I conceived the idea of typing those wealthier classmates' papers. With my new portable Smith Corona and the skills I had developed typing accident reports so many summers, I was able to drum up a well-paying clientele. Stunned by how illiterate my classmates were ("How did these guys get into Yale?" I asked Laurel, who wised me up to the advantages of having a father who had gone to Yale before you), I added the service of correcting my clients' grammar and making sure their arguments proved their points, which expanded my business exponentially.

I might have continued to slip twenties beneath my mattress forever, but I hated the way the services I provided made me feel inferior to my clients, performing distasteful tasks they didn't have the inclination to perform themselves, so I accepted the university's offer of a job preparing the compounds the pre-med students needed to identify in their labs.

Given my own performance in chemistry lab, the assignment seemed as wise as placing one of the Three Stooges in charge of the Smithsonian. Sterling Chem was a dismal brick castle; the basement where I had been told to meet my supervisor was as dreary as a dungeon. I expected to be greeted by a one-eyed hunchback named Igor. But the young man I found stooped over a bench, measuring powder on a scale, pulled up his goggles to reveal the kindest smile anyone could hope to see. Greg was a senior philosophy major from one of those midwestern states known for its abundance of wheat and corn and vowels; when he spoke, he sounded as if he were twanging on a saw. I could tell he wouldn't be impressed by my bumbling Lucille Ball act. He caught me squinting at some labels and asked if I needed glasses, and when I mumbled that I already owned a pair, Greg's disbelief that I would allow my vanity to interfere with my ability to read persuaded me to wear them. (Besides, if I didn't wear glasses, I would need to protect my eyes with goggles that were even dorkier.) Even as I continued to run amok in my own chemistry lab on the upper floors of Sterling, I managed not to break a single piece of equipment in the basement, nor did I prepare a solution incorrectly.

My hours in that dungeon were some of the happiest I spent at Yale. Greg talked to me while we worked—about Hume, Descartes, and Kant, and whether he should go on for a doctorate in philosophy or opt for a more practical career in computer science. Ashamed of never having heard of Hume, Descartes, or Kant, I resolved to sign up for a philosophy course so I could impress Greg. But I also felt something genuine stir inside me. Philosophy—of course. Here was the discipline in which the most important questions were placed at the center of the discourse rather than being shunted off to one side. How could any of us go on living, knowing that we had to die? How should we guide our actions if we didn't believe in God? Greg scoffed at Ayn Rand's insistence that selfishness could be a virtue, preferring Kant's categorical Do Unto Others. Once, he chided me for cutting from one sidewalk to the next; I pointed out that my footsteps hadn't damaged the grass, even as I knew that according to Kant's imperative, I should cut corners only if I wanted to live in a world in which everyone cut that same corner, in which case the grass would soon be dead. (All these years later, whenever I face the temptation to cut across a quad at the University of Michigan, where I teach, I remember Kant and take the longer and less convenient route on the concrete walk.)

In this way, Greg joined the long line of men upon whom I would have gladly bestowed my virginity, if only he had consented to accept it. He invited me to hike to the top of Sleeping Giant State Park, where we admired the view of Long Island Sound and shared a moment I was certain would result in Greg leaning down to kiss me. When he didn't, I tried to figure out what I was doing wrong. I had been told men wanted only one thing from a woman. So why didn't any man want that one thing from me? I fantasized all the time about having sex with men. I had never had a crush on another girl. But I didn't feel the least bit feminine. Surely other women didn't feel all this energy. They weren't as obsessed with winning. None of them seemed to want to study physics or math. And, my most shameful secret of all: other women got their periods.

In seventh grade, I had bought a sanitary napkin from the vending machine in the ladies' bathroom. The napkin in my purse grew stained with ink; the cotton padding grew brittle and decayed; the safety pins with which I was supposed to attach the napkin to my panties flaked to rust; and yet the "friend" whose visit I so keenly anticipated never did pay her call. I did what any self-protecting teenage girl would do: I lied. Once a month, I moaned about the cramps I wasn't having. Having eavesdropped on a session in which my sister tutored her friends in the fine art of inserting a tampon (put one foot on the toilet; dab Vaseline on the tampon's tip), I took my own friends into my grandmother's bathroom in the attic and tutored them.

Finally, the summer I turned seventeen, I got my period. But my period wasn't very periodical. It came, then vanished like some puppy that's been let out to do its business, too young and confused to find its way home. When my parents dropped me off at Yale, I wasn't surprised that my period didn't come along. Such a recent companion, it wasn't yet loyal. It couldn't pick out my scent amid the pipe smoke, shaving cream, and sweat suffusing the halls in which women had been living for such a short time. A theory then current dictated that shared schedules of feeding and light would cause a woman's period to fall in sync with the periods of the women she lived with. But after six months, I stopped by health services and asked to see a gynecologist. I peed in a Dixie Cup and allowed a nurse to take some blood. Then I was ushered into an office, where I took a seat across from the doctor.

"Do you mind if I ask a few questions?" Dr. M. said. "Are you proud of your abilities as an athlete?"

"Why, yes," I answered modestly.

He reached across the desk and stroked my face. "And this? Have you noticed that you have peach fuzz on your cheeks?"

Unlike several of my friends, who had been bleaching their lips since puberty, I prided myself on not having a mustache. But that didn't contradict the evidence that my cheeks were furred with a light blonde fuzz.

"You don't need to answer this," Dr. M. went on, "but I'm fairly certain you have hair on other parts of your body. Your belly, perhaps? Your breasts?"

If you had tortured me to reveal my most shameful secret, I might, if the pain had grown unbearable, have confessed that a faint line of dark hair, like an army of ants, traveled from my navel to my vagina. Thankfully, Dr. M. provided me with an out. *You don't need to answer*, as if the Constitution protected me from incriminating myself for the crime of being hirsute. I sat mutely and allowed him to infer that I was only a few hairs short of running away to join a sideshow. At that, Dr. M. leaned in and delivered his coup de grâce. "And you're good at science?" he asked. "And math?"

Why, yes, I admitted. I was the first female undergraduate to major in physics.

Well! he said. He intended to send away my blood and urine to be tested. But even without knowing the results, he could diagnose my problem.

Yes? I said. I was at that age when a woman suspects that a single freakish flaw prevents her from resembling other, more normal women, and if only someone would tell her what it was, maybe she could correct it.

"You've got too much male hormone. I wouldn't be surprised if you are sterile. You can take the pill in this envelope—a morning-after pill, it's called. It's a strong dose of female hormone. It might jolt your system into giving you a period. But it probably won't. When we get the results of your lab tests, we'll give you a call and you can come in and start your treatment."

I thanked him and left. That night, I told my mother I had too much male hormone and was therefore sterile. I couldn't figure out why she was so upset. I didn't want to have children anyway. I had gotten such a late start studying physics, I would need to spend the rest of my life catching up.

My mother remembers none of this. Maybe she suspected the diagnosis was ridiculous; she and my sister were more generously endowed than I was, but they, too, were late bloomers. I took the morning-after pill, and my period returned the morning after. The gynecologist's office never called. The following month I bled again, and the month after that, until, in my thirties, my period failed to come; as it turned out, I was so fertile I had gotten pregnant from a

single sperm sneaking through a pinhole in my diaphragm and surviving a toxic-waste-dump full of spermicide.

But all that came later. My first year at Yale, I was relieved to
solve the mystery as to why my period had come so late, why I loved
running and hitting balls, why I was so ambitious and competitive,
why I was more interested in science and math than other girls, and
why, try as I might, I couldn't give away my virginity. I wasn't a
woman like other women. I wasn't a woman at all. As it turned out,
I was a man.

CHAPTER SEVEN

Electricity and Magnetism

Returning home after my first year at Yale, I felt the way Hermione Granger must have felt leaving Hogwarts to spend the summer with her Muggle parents. (Like me, Hermione is the daughter of a dentist; in Hermione's case, two dentists.) A day after acing my physics final, I found myself back at the insurance company typing accident reports while listening to the secretaries complain about their husbands and trading sexual innuendos with the adjustors. When I wasn't working, I relaxed in my father's hammock reading *Lord of the Rings* or played tennis against a former teacher who worked as a pro at Brown's Hotel. Or I hung around with Barry, who had moved from his cottage on the lake to a house just past the reservoir.

No matter that he hadn't written; when I showed up at his door, Barry would take me in his arms and hug me, then, eyes shining, pull back and exclaim, "You look wonderful! Tell me everything you've done since I last saw you! How are your professors? What are the other kids like? Is everyone as brilliant as you thought they'd be?" And out everything came, in stories that made Barry gasp: "You're *shitting* me. That didn't *really* happen!" He would cook his latest favorite dish, or croon along with the soundtrack from *A Chorus Line*, sashaying across the floor as he motioned in my direction: *One . . .*

sing-u-lar sen-sa-tion . . . ev-ery little step she takes. One . . . thrilling combination . . . ev-ery move that she makes. . . .

I wanted never to leave that house. I would stay home and marry Barry, play tennis, read novels, rub garlic around the salad bowl while he prepared the crepes ("The first pancake needs to be sacrificed to the Crepe Goddess," he would say, tossing it in the sink), after which we would open a bottle of wine, snuggle on the couch, and listen to Jacques Brel or Sondheim. But when September rolled around, Barry made no move to hold me back. And really, once you have enjoyed the thrill of taking the train to Hogwarts, how can you resist climbing on that train for another term?

Silliman is a four-sided castle with arched, gated entryways and crenellated towers festooned with gargoyles. Instead of using a Sorting Hat to determine who roomed where, each residential college held a lottery. Laurel and I decided to stick together, but we needed another two girls to share a suite. Erika and her roommate Sherry had teamed up with an upperclasswoman, let's call her Rita, who, by reason of her seniority, had earned the right to live with four female Sillimanders of her choosing in an eccentrically large suite called the Penthouse. Originally, the Penthouse had consisted of one large bedroom with a fireplace, one smaller room to accommodate that student's servant, a living room, and a smaller room for another student, but previous generations had built additional rooms into the recesses of the castle, with room after windowless room fanning out from the original suite. I wondered why no one in Rita's own class wanted to take advantage of the opportunity. But she promised that Laurel and I could share the double room with the fireplace, and we were excited by the prospect of living in a space as large as our entire suite the year before.

At first, we enjoyed our spacious new digs. Then I noticed that whenever I mentioned where I was living, the boys in the college gave me strange looks. I couldn't imagine why until someone confided that Rita had taken it as her mission to go through the Silliman directory and have sex with every guy from A to Z, thus explaining the steady stream of men who visited her.

Then Rita broke off her relationship with the one guy in Silliman who really did seem to love her. Laurel and I were awakened by a pounding on our door, then loud noises from the back rooms of the Penthouse. In the morning, we discovered the living room had been trashed and several benches in the courtyard tossed about as if by an angry giant. After that, we kept to ourselves, barricading ourselves in our room and avoiding Rita.

Looking back, I can't help but wonder why I didn't go out that night to make sure Rita was okay. It never occurred to me that she might be as fucked up as I was, both of us longing for male approval, resorting to whatever means we could use to get it. Maybe we had more in common than I thought.

A few weeks into term, I tried out for the tennis team. Playing a sport would mean I wouldn't have time to continue working for the Chemistry Department, but the job had lost its appeal now that Greg had left New Haven (and yes, he went on in computer science and not philosophy). And with all the time I spent sitting in a small room solving problem set after problem set, I needed the sheer physical release that pounding a ball provided. Impressed by how quickly I covered the court and how fierce I was at net, the coach awarded me a spot on the varsity team, only to watch in amazement as I lost to far weaker players, slipping lower and lower until I was clinging to the bottom of the JV ladder.

Still, I loved playing the game. And I can't help but think that if physics hadn't been so all-consuming, I might have remained healthy and sane. I might have discovered a way to feel more comfortable competing on the tennis court without being afraid that playing my best would make my teammates hate me, which might have enabled me to feel more comfortable performing to my best ability in physics and math without caring what my classmates thought.

Also, those afternoons on the tennis court remain in my mind as the last time I enjoyed anything resembling a normal college life. Just as I was catching up with the boys in my class, I was offered the chance to graduate a year early, and I had no time to enjoy anything as frivolous as playing tennis.

"Dear Ms. Pollack," the letter read. "I am writing to report that at the conclusion of your Freshman year, the Registrar's Office reports that you have received a total of three or more College Credits. . . . If you wish to complete the requirements for the bachelor's degree in six rather than eight semesters, you should consult the Director of Undergraduate Studies in your intended major."

Ironically, the four AP exams I had taken to compensate for the weak curriculum at my high school, added to the extra courses I had struggled through freshman year to make up for my deficient background, had left me with so many credits that the Dean was inviting me to accelerate. Having been denied the chance to skip a grade in elementary school, and again in junior high, I couldn't resist the offer. And how could I turn down the opportunity to save my father eight thousand dollars in tuition? I scheduled a meeting with Professor Parker, who said all I needed to do to fulfill the requirements for the major was to take one extra physics course every term.

Which is how I came to be enrolled not only in the third semester of introductory physics and the first term of its accompanying lab, but also Linear Algebra, Statistical Thermodynamics, Physiological Psychology, and a reading course in German. At first, I managed to keep my head above water. I loved linear algebra, in which complex physical systems are reduced to arrays of numbers that are manipulated to predict how the systems will behave—it's as if the meaning of life could be reduced to a Sudoku game and solved. The instructor was the youngest full professor ever hired at the university. Reputed to be an avid squash player, he seemed too tall to fit inside a squash court, and his way of floating in a languid daze didn't seem suited to the game's lightning-fast moves. In physics class, when Professor Zeller explained how the forces on a spinning tire caused the wheel to torque, he would transfix you with his gaze; in Linear Algebra, Professor Howe directed his dark, liquid eyes on some point beyond the ceiling, envisioning whatever abstract principle he was trying to explain and speaking in the unearthly cadences of a medium. Like Zeller, Roger Howe was an attractive man, his looks marred only by the way his two front teeth folded over each other, although by the end of Linear Algebra, I found even his crooked smile endearing.

The next semester, I signed up to take Applied Calculus for no reason except Professor Howe was teaching it. When the course turned out to be an exploration of the phenomenon that had been obsessing me for years—the ability of a ray of light to sniff out the quickest path, or a soap bubble to decide which shape to take—I felt like a disciple who has found the master she has been seeking all her life.

In the meantime, I needed to get started on the required three-term sequence in quantum mechanics. The professor, a young Hungarian named Peter Nemethy, was so charming I began to wonder if Yale had cornered the market on handsome young male physicists. Perhaps the university went out of its way to recruit charismatic physicists as a strategy for attracting women to the major, along with preppy men, who could rest assured that a career in physics needn't exclude them from membership at the club or the ministrations of the female sex. Let MIT hire the genius misfits; Yale would settle for the merely brilliant who happened to be presentable. Although Professor Parker was older and more restrained than Professors Nemethy, Howe, and Zeller, he was no slouch in the looks department, and he had the added allure of his Spider-Man mystique; it was if I were being taught by my own private quartet of superheroes.

Of the four, Nemethy seemed the least approachable, aloof in the tradition of Old World professorial aloofness. He called on no one by name and accepted no excuses for late homework. When his fingernails scratched the board—which they often did, he wrote in such a flurry of excitement—he didn't flinch, even as his students writhed. "What are the guts of the physics of the thing?" he would ask, punching a chalked equation with the underside of his fist. When no one answered, he took two long steps sideways, folded one leg behind him, and pressed the sole of a scuffed moccasin against the wall. Levering with his hands, he would propel himself forward, then allow himself to fall back against the wall, forward, then back, until, with a mighty heave, he propelled himself across the room, arms and legs flailing as if they were jointed with pipe cleaners, and I couldn't help but notice he had left his footprint on the wall, overlapping with the other footprints he had imprinted there before. I could have listened to the thick

Hungarian rumble of his voice for hours. For two semesters, I thought he was saying, "Let's devil up the math!" rather than *develop*. "Look at this equation! That is quite neat, don't you think, what we have deviled up? No, no, it's very neat! Such a gross approximation we have used, and look! We can predict the lifetime of an alpha decay!"

The boys weren't nearly as enamored with Professor Nemethy as I was. They grumbled when he gave exams, which covered not problems from the book, but topics gleaned from research he and his colleagues were carrying out at Brookhaven on Long Island, Fermilab in Chicago, or, in Professor Nemethy's case, the giant collider at CERN, the European Organization for Nuclear Research. I grumbled, too, but only to camouflage that, once again, I was bringing up the curve. I loved everything about that course. Here at last were the miracles I had glimpsed in those archaic monographs I had taken out from the community college in junior high. Every particle could be described by the wave-shaped function $\Psi(x,t)$, which predicted the probability of finding that particle at any position x at any time t. (In this magical new universe, it was impossible to know where anything was, only where it might be.) A particle confined to an infinitely deep potential well had an infinitesimally small but very real opportunity to escape by tunneling, which was like saying a convict could escape from a prison that was impossible to escape from by attempting to escape infinitely many times. Listening to our professor describe what happened when an electron collided with a positron reminded me of listening to Mr. Spock explain particle-antiparticle annihilation to Captain Kirk. I never summoned up the nerve to talk to Professor Nemethy in his office. But I could sense a connection between us, as if the fire in his eyes ignited a sympathetic flash in mine. Far from minding when I interrupted a derivation, he seemed delighted that I was following what he said. He treated me no differently than he treated the men.

What he didn't realize was that that every time I asked a question, I missed half of what he said because I could hear my classmates think that if I weren't so dense, we could cover twice as much material. Like Professors Zeller, Parker, and Howe, Professor Nemethy had no idea how desperately I needed his encouragement in ways his other students didn't. He didn't understand that the list of eighty-eight male and two female Nobel Prize winners printed on the inner covers of our book, combined with the absence of a single female faculty member in physics or math and the score of 32 that I had earned on my first exam, made me feel as if I had no right to be in his class.

The best—or worst—example of this phenomenon came in Statistical Thermodynamics. Most physics majors regard thermo as an outmoded branch of science that concerns itself with the behavior of gases and steam engines. But the course appealed to my curiosity about abstract questions, such as the impossibility of building a perpetual-motion machine and the reasons everything moves from a state of less to more disorder. By all rights, I should have gotten along with the professor. An avuncular New York Jew, Professor K. had the huge nose, brushy white mustache, bulbous head, extravagant eyebrows, oversized glasses, and crinkly, kind eyes I associated with the more intelligent guests at my family's hotel (not that any of the guests at Pollack's had graduated from Columbia at eighteen, earned a doctorate in physics from MIT, or made a name for himself as a theoretician before editing the collected papers of Albert Einstein). In another world, I might have found a mentor in Professor K. But in this world—the world in which I showed up to the first class to find myself one of only two female students—Professor K. and I got off to an unfortunate start. Or rather, he got off to an unfortunate start with the other woman, and I took this to reflect on me.

"Would anyone mind if we were to schedule a problem session one night a week after dinner?" Professor K. asked. None of the men objected. But the other woman, a graduate student in astronomy, told Professor K. that an after-dinner session would interfere with her ballet classes. Startled, the professor asked if she couldn't miss her dance class one night a week. And the student told him no, she had signed up for Statistical Thermodynamics thinking it would meet in the afternoon; he couldn't just reschedule it for the evening and expect her to miss ballet.

The professor's eyebrows shot above his glasses. The male students fidgeted and rolled their eyes, and even though I would have canceled an engagement to dance at Lincoln Center rather than miss a physics class, I slunk lower in my chair. Very well, the professor said. Instead of meeting after dinner, the problem session would be scheduled *during* dinner, which clearly would inconvenience Professor K. and the rest of us.

Even worse, when we met for that first session, the astronomy student didn't show up. For whatever reason—a scheduling conflict, an unwillingness to take a course from a professor with whom she had had a run-in—the ballerina dropped Statistical Thermodynamics, leaving me to assume that every man in the room, the professor included, would take his resentment out on me.

The students in that class were a sullen, unwashed bunch. By some secret agreement, they had decided to convene early for each session and compare their answers to the problem sets, but no one had invited me. I suppose I felt hurt. But I no longer expected the men in my classes to consider me one of them. It was as if we were on opposing teams, and they had no reason to help me win. At first, I didn't dare to stop Professor K. to ask a question for fear of giving him further grounds to think all female students were difficult or demanding. But I hated not being able to understand the principles that underlay the first or second law of thermodynamics, so I ventured to ask him to slow down just a bit. He grew visibly irritated, slackened his pace, then returned to normal speed. For all I know, if a male student had asked him to slow down, Professor K. would have responded irritably to him as well. But no male student did.

The one time I stayed after class to ask our professor to explain a theorem, I could sense he was annoyed. Or maybe that was only my imagination, as it might have been my imagination that every time he handed me back a problem set, my grade was lower than I had predicted. Why had he docked me so many points for making a simple mistake in math? Five points, okay, but ten? I studied the material until I was confident I could solve any problem Professor K. might dish out. But again, my score on the final was lower than I expected. Gathering my courage, I asked if he realized I had missed an A- in his course by only a few points, and would he be willing to look at the exam again to see if he had marked it fairly?

As a teacher myself now, I understand why he lost his temper. How dare I consider it *his* fault that I received a lower grade than I "expected"? If I had *earned* a few more points, I would have gotten them! The encounter was so unsettling I decided I would never again intimate to a professor that being a woman had affected my grade. And I made a note that there was no place in my chosen field for anyone who dared to put ballet on a par with physics.

As it turned out, the A that I was denied in Statistical Thermodynamics was more than balanced by the A that I didn't deserve in Electricity and Magnetism. I was supremely uninterested in how to calculate the shape of the electromagnetic field generated by an oscillating electric dipole. The professor didn't understand the subject any better than I did, and even if he had, he wouldn't have been able to convey what little he knew. In his three decades on the faculty, Professor C. had neither published nor perished; rumor had it he had done no research since completing his doctoral thesis in 1949. He had come to Yale when the school needed physicists and had gotten tenure before the competition for positions grew intense, but he had never been promoted. The one time I visited his office, I saw affixed to the door a hand-lettered card that read, "The only thing of which I'm certain is that I make mistakes," a sentiment I might have found endearing if I hadn't been depending on the occupant of that office to steer our class through a stormy sea of electromagnetic waves.

A short penguin of a man, he waddled into our room that first day in a dark blue suit and white shirt, hands behind his back, head bent, as if he carried an egg between his shoes. Immediately, he began pecking at the board with the tiniest bit of chalk, dotting out equations too small to be read. Every class for the next two months, he worked his way through that same set of equations, the shoulders of his suit growing heavier with chalk dust, the sleeves growing more and more smudged with white. There were only five of us in the class, but he never addressed a remark to anyone. A sour-looking graduate student in the front row sat with a pen between his teeth, occasionally shouting the professor's name, less because he had a question than he wanted to bait the poor man. "Mr. C.!" he would cry. "Mr. C.! Can you show how you derived that formula?," at which Professor C. would jerk around, put his knuckle to his mouth, and tear his fist across his teeth as he attempted to think of an answer.

I would have found the class unbearable if not for Fitz, a scruffy but sweet physics major I knew from Silliman. No one understood E&M, Fitz assured me. Stick with him, he said, and we would muddle through together.

"Really?" I said. He would help me get through this disaster of a class?

"Sure," he said. "But first I have to take care of that jackass in the front row."

"Yeah?" I said. "What are you going to do?"

"I don't know, I'll think of something."

The next time we met, the graduate student asked another question, and Professor C. spent another forty-five minutes tearing at his knuckles. As we were walking out the door, Fitz, who gave off the *don't fuck with me* aura of someone who idolized Hunter Thompson, flicked open his penknife and pressed it against the graduate student's hip. "Pull that again, sonny boy," he hissed, "and I'll cut off your balls."

After that, Fitz and I spent the semester sitting quietly in our seats and working out the problem sets as Professor C. chalked equations on the board. I didn't understand any more about electromagnetic fields at the end of term than I did the day it started, but I still received an A. For all I know, everyone received an A. (I didn't ask Fitz, but he must have done well enough, because he went on to become a doctor. In all my four years at Yale, he was one of only two male physics students who treated me not as an oddity, but as a friend.)

What I didn't know about electricity had no effect on the grade I received in E&M, but that ignorance nearly killed me in physics lab. Given my performance the year before in chemistry, I would have been nervous no matter what. But everything conspired to make that physics lab even more of a disaster than it otherwise might have been. The director hustled in like a dour Groucho Marx, barked an explanation of what we were supposed to do, then vanished, leaving us with nothing but the stink from his cigar. I was the only woman in the lab, although I felt fortunate in having as my partner a brawny chemical engineer named Al. I found myself hoping Al would ask me out; when he didn't, I consoled myself that I had snagged a partner who knew what he was doing and didn't seem to mind I was a girl. I felt comfortable in his presence—he reminded me of the working-class Polish Catholic guys I knew from home, an impression reinforced by the cast he came in wearing on his hand one day.

"Al!" I said. "What happened?"

"Bar fight," he said, smiling in a way that made me think the other guy had come off worse.

And yet, Al was no more successful at getting our experiments to work than I was. *This equipment sucks,* he muttered. *The reason we can't get the right results is that most of this junk is broken.* We would grab the lab director to ask for help, but the man would wave his cigar and tell us the equipment was fine, we weren't following the write-ups. One day, when we were supposed to be hooking up an oscilloscope to some other piece of equipment, he waltzed in and made a snide comment about how we had connected everything wrong.

"It's *not* wrong," Al said. "I've checked and rechecked the diagrams. The problem is the equipment doesn't work."

"I hardly think that's likely." The director stuck his cigar between his teeth and yanked wires off electrodes, then clipped each wire to something else. "There," he said. "But I shouldn't have needed to do this for you."

I was less certain than Al that the problem lay with the equipment. But I was anxious to complete the measurements before the dining halls closed. I reached out to position the oscilloscope, but when my fingers made contact with the metal case, a shock threw me across the room. When I came to, I was sitting on the floor. Al rushed over and asked if I was all right. The director came in and mocked us. With clenched fists, Al intimated the lab director must have connected the power source to the casing of the oscilloscope.

"You're blaming me?" He waved his cigar in Al's face. "Let me tell you something. I had everything connected right when I left this room. You must have moved something and screwed it up."

Al exploded with a voltage higher than the oscilloscope. "We didn't touch a thing! You nearly got my partner killed!" Al pulled back his arm; I grabbed his bicep. No punch was ever thrown, but the director threatened to have Al expelled. He relented, but from then on refused to acknowledge either of us. Al and I blundered through the rest of the experiments, then wrote up the results as extensively as we could. I didn't ask Al about his grade, but when I saw that mine was a C, I found the director and accused him of lowering my grade because he was angry at my partner. I wish I hadn't betrayed Al that way. But I was sure a C in physics lab would doom my career.

The director, whose dingy office reeked of cigar smoke, jumped up from his desk. "Get out of here before I lower it to an F!" I took the hint, only to find out later that he had given in and raised my grade to a B.

Not surprisingly, I entered the second term of physics lab convinced that even the most innocuous piece of equipment was out to kill me. This time, we each worked on our own, moving from dungeon to attic, encountering in each dismal cell a Madame Tussaud re-creation of one of the triumphs of experimental physics. But even when I was presented with the same apparatus Robert Millikan had used to calculate the charge on an electron, I couldn't reproduce his results.

The final requirement was that each of us replicate a famous physics experiment that hadn't been covered in the first part of the semester. Tempted as I was to thumb through my parents' *World Book Encyclopedia*, I consulted the considerably more advanced physics reference section in Kline. Most of the experiments required a nuclear reactor to carry out. But I finally came across a description of something called Chladni plates, which had been invented in the 1700s by a German physicist and musician named Ernst Chladni, who was interested in a phenomenon called "standing waves."

Everyone is familiar with standing waves. Pluck a rubber band and sound waves will travel along its length, then get reflected back. The waves traveling to the left will have their crests and troughs in different places than the waves traveling to the right, which means most of the crests and troughs will cancel out. But for just the right length wave, on just the right length rubber band, the crests and troughs traveling to the left will coincide with the crests and troughs traveling to the right, resulting in a wave that appears to be standing still. For any particular rubber band, only a few waves will fit perfectly along its length, and for those few frequencies, the rubber band will vibrate until friction bleeds away the energy, which is why, if you pluck a rubber band of a particular length, it will sustain only certain musical tones (and their harmonics).

A similar principle governs the two-dimensional waves that can be sustained on the surface of a drum, or the three-dimensional sound waves that can resonate inside a barrel. I was fascinated by standing waves, less because I cared about designing a violin than because standing waves seemed to underlie everything in the universe, from vibrating guitar-strings to the orbits of electrons to the ability of light to sniff out the shortest path from here to there. I decided I would replicate the experiments in which Chladni had created visible standing waves by sprinkling sand on a thin metal plate he stroked with a rosined bow.

Just as there are now websites purporting to guide students through science fair projects in electroplating and hydroponics, a variety of sites demonstrate how to generate Chladni figures. (My favorite shows a barefoot young woman singing notes of unearthly beauty into a microphone connected to an illuminated plastic sheet on which a layer of salt vibrates to form eerily sensuous patterns.) If I had been a music head like many of the boys in physics, I could have figured out how to use a discarded speaker to carry out the experiment; lacking such expertise, I decided to replicate Chladni's old-fashioned methods. I don't recall where I obtained the thin metal plate, or how I balanced it on a needle; I only remember sprinkling sand on my creation and scraping it with a bow for hours—faster, slower, until finally I persuaded a few grains of sand to dance.

I was faced with the dilemma of whether to admit my failures or lie and pretend I had come up with the results I meant to achieve. Already, my propensity to commit scientific fraud was becoming a trend. First, the Incident of the Orange Cis-Boom-Bah. Now, the Case of the Counterfeit Chladni Plates. Should I be honest and risk failing, or should I write up the Millikan oil-drop experiment as if I had derived a reasonable value for the charge on an electron? Should I admit I hadn't been able to reproduce a Chladni pattern, or should I draw variations of the diagrams that accompanied the article in the encyclopedia?

Reader: I cheated. Whoever graded my lab reports didn't notice. Or maybe every generation of physics students knows that, like generations before them, they are expected to fudge their results. Reader: I got my A. But my guilt was such that I vowed never to step foot in a lab again. I would make it as a theoretician, or I wouldn't be a physicist at all.

The pressure to graduate in three years instead of four, combined with my loneliness and isolation from being the only woman in my classes, strained my nervous system to the limit. One afternoon, I climbed to the top floor of the math building to use the only women's restroom, and on the way down I missed my footing. I grabbed the railing in time to keep from tumbling to the bottom. But in my mind's eye, I saw myself cracking my head against every marble step. After that, each time I walked down a flight of stairs, I needed to grip the rail and inch my way down. When I crossed a street, I pictured my skull getting crushed by the tires of a car. I was like a young pitcher who becomes obsessed with what might happen if he hurts his arm; if something injured my brain, I would lose my access to the world I so desperately hoped to enter. But understanding the cause of my obsession didn't help me to fight off the phobia.

Around this same time, I lost my ability to sit still. As Professor C. stood pecking on the board, I grew so agitated I needed to hurry out. In the evenings, I might slip into Woolsey Hall to catch a performance by a visiting symphony, or the Yale Russian Chorus, whose haunting a capella ballads brought me to tears. But after half an hour, I started to twitch. We had learned in Statistical Thermodynamics that if you increased the pressure on a gas, the molecules collided against the walls until the gas exploded, and that was how I felt, as if my molecules were vibrating so fast I needed to hurry to my room to work on my problem sets or risk exploding.

The only time I seemed able to relax was during meals. I ate and ate, and every time I went home, my family told me that I was getting fat. Unable to stop, I invented what I thought was an original solution to my dilemma, not realizing that all over America, other women my age were inventing this same solution, and all of us with eating disorders would come to represent an epidemic so widespread, magazines like *Newsweek* and *Time* would feature us on their covers.

Hearing other women say smoking cigarettes helped them keep down their weight, I started bumming cigarettes. The tactic worked, until I realized that if I continued to smoke, I would need to buy my own pack. As a compromise, I bought tobacco and a pipe, figuring I wasn't yet eccentric enough to walk around campus puffing on a pipe, and so would smoke only in my room.

Then again, it was just a matter of time. Brooding on my diagnosis from the gynecologist, I began to think of myself as a man. Not as a lesbian—I had never had a crush on another girl—but a man in a woman's body. Or a woman in a man's body. I found a fedora in a thrift shop and began wearing it all the time. I even wore it home for winter break.

"Don't you dare go out in that hat!" my father ordered. Startled, I asked why not. "That's a man's hat," he snapped. "I don't want the neighbors thinking my daughter is Marlene Dietrich." I didn't know who Marlene Dietrich was. I only knew my father didn't want our neighbors to whisper that his daughter was so perverted as to be walking around the block in a fedora hat.

All of this—the phobias, the inability to sit still, the habit of forcing myself to throw up after every meal, the sense that I was a man in a woman's body—led me to yet another obsessive fear: that I might end up locked in an institution. I signed up for Physiological Psychology in part because I wanted to explore the question that had interested me as a kid—how human consciousness came to be—and partly because I was interested in figuring out how a person's neurons might misfire in such a way that she couldn't walk down the stairs without spiraling into panic or repeating the same obsessive phrases in her head.

A few weeks into term, the professor—one of only two female instructors I had at Yale—handed us each a sheep brain encased in plastic, along with scalpels and instructions on how to perform the required dissection. To avoid stinking up the room I went up to the roof, where, as a thunderstorm brewed over Long Island Sound, I jabbed my scalpel through the Seal-a-Meal bag and stood with another creature's brain nestled in my palm. How rubbery it felt, how smooth, like a cauliflower left on the counter too long; the cortex was light gray, but when you sliced off a segment, the floret inside was white. Following the instructions, I split the brain from end to end, noting the corpus callosum, the medulla, the pons. I had no desire to be a doctor. But I wanted to understand how the human brain worked.

Most of the paper I wrote that term reads as if I had stitched it together from the cadavers of other texts. But the prose comes alive when I describe the first computer program sophisticated enough to simulate the firing of human neurons. When the model's creator applied a stimulus, the lights pulsed on and off in a "roughly periodic function," which is another way to describe a standing wave. Maybe standing waves could be excited in a neural net, I theorized, just as they could be excited in a violin string. Rather than dying out, the stimulus might produce a wave that would propagate throughout the neural net and set off standing waves in any other nets to which it might be coupled.

Even more exciting: a neuropsychologist named Karl Pribram had noticed the similarities between a new photographic technique called holography and the process by which memories are stored in the brain. In holography, an image is captured on a filter until light of the same interference pattern is allowed to hit it, at which the image is released. According to Pribram, the neurons in a brain act as a filter for a three-dimensional hologram, storing patterns of input from receptors. "The body is as complex as the physical universe," I wrote, "and just as early physicists had to abandon their concrete model of the atom for the unpicturable world of wave mechanics, physiologists must explore more revolutionary and theoretical languages for explaining perception, consciousness and memory."

Although the professor was right that my paper would have been stronger if I had compared the predictions of the mathematical models to the way actual neurons work, I can hear in that essay the excitement of a young scientist who has glimpsed the future to which her many interests may be leading. After that course, I was determined to find out more about the startling new fields of artificial intelligence and information theory. To do so, I would need to learn to program a computer—a machine with which I had so little familiarity that, in the essay I had just turned in, I kept spelling its name *computor*.

As it happened, my chance to develop a more intimate relationship with Yale's new PDP-10 and IBM-370 mainframes coincided with my determination to develop a more intimate relationship with Michael Zeller. Because he taught the introductory sequence, I no longer had the opportunity to see him on a regular basis, let alone impress him with how far I had come from the shaky freshman who had nearly flunked his class. As if I didn't have enough to keep me busy, I stopped by and asked if he needed an assistant. He carried out his research at Brookhaven, but he was, he said, eager to use Yale's new computing facilities to analyze his data. He had been looking for a graduate student who could write a program that would simulate a collision between a K-meson and a stationary proton and allow him to predict the locations of the particles that would be created.

I couldn't believe my luck. I didn't have time to take a course in computer science—if I wanted to accept Professor Zeller's mission, I would need to teach myself. I went straight from his office to the basement of Kline Library and checked out the two manuals I found on FORTRAN. The frustrating part wasn't figuring out how to write a program; it was standing in line to use one of the few machines to punch your program onto cards, then standing in another line to use one of the few machines that read them, then going home and praying you hadn't written a line of code that trapped the computer in an endless do-loop. Monitors didn't exist; the only way to model a simulation was to ask the computer to print out a series of graphs on which the positions of the particles were marked by Xs and Os. Months later, just as the computer time in Professor Zeller's account was running out, I picked up the latest version of my program and realized with a whoop that the simulation had finally worked. How cool was that? A bunch of Xs and Os generated by a machine had predicted the paths that some barely existent particles would take after a collision between two other particles no one had ever seen! Here was research that didn't require glassware, acid, vacuum pumps, filters, oscilloscopes, generators, soldering irons, saws, or electric drills. I didn't look down on experimentalists for dirtying their hands with particles. But I preferred to spend my life thinking about the results such physicists might obtain.

The last day of sophomore year, the professor who taught my course in Differential Equations asked me to stay after class. I had no reason to be concerned. I was the only woman among dozens of engineering majors—the version of the course offered in the Math Department had conflicted with my schedule—but I hadn't lost a point on any of the exams. Still, I wasn't expecting Professor Alben to say he had never had a student do as well as I had. Perhaps I would be interested in working in his lab that summer? He needed an assistant to carry out a series of experiments growing crystals.

Crystals! I knew all about growing crystals! Hadn't I grown crystals in elementary school? As Professor Alben led me around the lab, explaining his experiments and adding tantalizing tidbits about superconductors and solar-powered energy, my dreams for the summer crystallized in my head. But when I called home to tell my parents, they wouldn't hear of their daughter living on her own, especially in a dangerous city like New Haven, so, unlike my male classmates, who would be working internships in their fields, I had to turn down the job.

And suddenly I was back home in my childhood bedroom, with its pink walls, cherry-red carpet, and frilly white polyester quilt. Sequestered in such surroundings, I ought to have been able to relax. And yet, being cooped up in a pink-walled, cherry-carpeted room is hardly a cure for nerves. To escape the house, I drove out to see Barry. He was as kind and loving as ever. But I was beginning to sense that whatever sexual-magnetic field this man was throwing off wasn't being emitted for my benefit.

Finally, Barry called and said he had something to discuss. I asked if I should come to his house, but he said no, I should meet him at a bar; without a drink, he might not find the courage to say what needed to be said. This is how innocent I still was: I thought he was hoping to find the courage to make good on his proposal. I thought he was asking me to set the date.

When I got to the bar, he ordered us each a drink—scotch for him, tequila sunrise for me—and before either of us could take a sip, he blurted that he was gay. He was sorry, he said. But he couldn't admit that he was gay, even to himself. And now, with this horrible campaign by Anita Bryant to fire every gay teacher in America . . .

Later, I would think of a million questions I wanted to ask. What had it been like to grow up as a minister's son in the fifties? Had he ever been in love with a man? But I found the presence of mind to ask him only this: "If you knew you were gay, why did you propose?"

To which he offered the only reply that could help to heal the pain I had been suffering all those years. "Wishful thinking," he said. "If I could have fallen in love with any woman, I would have fallen in love with you."

CHAPTER EIGHT

The Philosophy of Existence

My first week back, I took a break to attend a mixer, drank more alcohol than I was accustomed to drinking, met a guy who was too drunk to care what subject I might be studying, went back with him to his loft, and lost my virginity to a classmate whose name I can't recall and can't bring myself to figure out by paging through the yearbook. Laurel was studying in Paris and I missed her. But as a junior who had accelerated into the senior class, I was entitled to one of the nicest singles in the college—or rather, a luxurious double, with an Indian-print sheet dividing my half of the room from the half occupied by an engineering student whose ambiguously gendered name protected me from admitting to my parents that my roommate was a boy. As unconventional as my suite mates had been the year before, that's how blissfully normal Lee turned out to be. The seniors on our floor behaved in such ordinary ways that I lived in fear they would discover *my* oddities, so I stowed away my pipe and fedora hat. When a bulb on my ceiling blew out, Lee summoned his much taller friend Stone Phillips to change it for me. Stone, who went on to become an anchor for NBC's *Nightly News*, was famous even then as an insanely good-looking philosophy major who doubled as the football team's starting quarterback. As he screwed in a new bulb, I kept thinking he would sniff out how peculiar I was, but he smiled pleasantly and invited me to accompany him across the hall for an impromptu party.

If that weren't enough to prove I had never been as cursed as I thought as I was, I noticed that one of my classmates in The Philosophy of Existence—a course whose title I couldn't resist—kept choosing to sit beside me. Ed was a Physics and Philosophy major a year older than I was, solidly built and square jawed as a lumberjack, if any lumberjacks had grown up on the Main Line in Philadelphia. The course was taught by a professor with a snowy Talmudic beard. At the start of each lecture, he would place his briefcase on the table, remove a pad, a pen, a watch, and then, after he finished lecturing, return each item to the case. One day, as he was performing this ritual, Ed leaned over and whispered, "I keep thinking he's going to take off his beard and put that in his briefcase, too." I laughed so hard I nearly choked.

Ed invited me to a movie. We went out for a drink and talked about Martin Buber, whom we had studied in class, until the evening ended with the two of us attempting to establish an I-and-Thou relationship with a tree, then the tree looking on in approval as Ed and I made out. He had attended a Quaker high school, but he didn't seem embarrassed to be a Jew, and one afternoon he took me to hear the Jewish chaplain lead a discussion about mysticism and the kabbalah. The rabbi asked who among us believed life was about reaching a desired goal, and I raised my hand. Then he asked who believed the journey was more important, and Ed raised his hand. Really? I said. You don't want to be successful? Sure, Ed said, but I want to be happy along the way.

It hadn't occurred to me someone might decline to pursue a goal in favor of being happy; even if a person had been miserable all her life, I was sure once she achieved success, her happiness would bleed backward to cast a rosy hue on everything that came before. That night, as Ed and I made love in front of the fire in my fireplace, I experimented with being content being exactly who I was, doing exactly what I was doing. And for once, I achieved the experimental result I had been hoping to achieve.

．　．　．

Even as I signed up to take the classes that would allow me to graduate a year early, I began to doubt I wanted to spend my life working so hard I couldn't walk down a flight of stairs or cross a street without hallucinating my skull might get crushed. Classical Mechanics with Professor Parker and the second semester of Quantum Mechanics with Professor Nemethy posed no serious threats to my sanity. But my math class, Real Analysis, turned out to be impenetrable. I could barely understand what real analysis was, let alone how to do it.

The course was taught by a courtly Japanese mathematician named Shizuo Kakutani, who was already acclaimed for his work in probability theory and would gain additional renown as the father of the senior book critic of the *New York Times*. That semester, we would be studying the numbers between zero and one. In fact, we might not make it further than the first chapter in our book. "Is like being tourist in foreign country," our professor told us slyly. "Some people, they go to Paris, they get on tour bus, hurry, hurry, hurry from one attraction to the next. They see the Eiffel Tower go by. Oops, there goes the Louvre! These people think they know Paris. Other people sit in same café all day, watch Paris go by them." He grinned. "*These* are the people who truly know Paris."

Professor Kakutani cared less that we be able to solve any particular problem than we grasp the essential beauty of real analysis. He told us that growing up in Japan, he had been pressured to pursue a degree in literature. He presented the diploma to his father. (Here, our professor mimed a bow and handed an imaginary scroll to his honored parent.) Then he asked permission to go on for a PhD in math. As our professor told his story, I thought how liberating it would be to move in the opposite direction—to prove you were capable of completing a degree in science, then spend the rest of your life reading and writing literature.

After three or four weeks of struggling through the proofs, I went to see Professor Kakutani to get his signature to drop the course. Oh, no! he said. You mustn't drop Real Analysis! If I stayed, he would guarantee I received a B. If I came to his office, he would tutor me all I needed. He even offered to come to my dorm and tutor me there. Embarrassed by his generosity and unable to admit that I didn't care about the numbers between zero and one nearly as much as he did, I blurted, "I couldn't possibly let you do that!," then obtained his reluctant signature and hurried out.

Years later, my sister's college roommate, Marie, who had earned her PhD in math and gone on to a successful career studying commutative algebra, told me real analysis is a killer and plenty of mathematicians never got the hang of it. Really? I said. And they still go on to become mathematicians? "Of course," Marie said. "That's so typical of the way that women think. Everyone has different strengths. If men don't do well in a course, they blame the professor. Or they bluff their way through and choose a specialty that allows them to avoid that subject. Women blame themselves. With women, it's all or nothing. They need to be perfect or they quit."

That same term, I stopped by Professor Nemethy's office to see if he could advise me on how to prepare for a career in theoretical physics. As a junior with senior status, I ought to be acquiring letters of recommendation from professors in whose courses I had excelled. Professor Nemethy, who had been sitting with his feet jammed against his desk, jumped up like a jack-in-the-box. He had been planning on getting lunch. Perhaps I would care to join him?

I had never eaten with a professor. Sure! I said, and we took the elevator to the cafeteria on the top floor of Kline. Despite the spectacular view, few undergraduates ate there, and most of the tables were occupied by lone male physicists and biologists who sat reading articles or scribbling equations as they ate. As I watched the scientists come out of the buffet and choose their seats, I was reminded of the Pauli exclusion principle, which dictates that each new electron is required to take its place in the next empty orbital before deigning to partner up in an already occupied shell.

We found a table, two electrons pairing up of our own volition, but for the first time since arriving at Yale, I was too excited to eat.

"So!" my professor said. "You must tell me why you are in such a hurry to become a physicist."

As much as I wish at least one professor had encouraged me to go on in physics, I can't fault Peter Nemethy for not being that one. Why would he have thought I needed to be *encouraged*? I was so gung ho that I was graduating a year early. If I had had me for a student, I would have tried to discourage me from graduating early, too.

"Why rush?" he said. "Aren't there other things you like to do besides physics?"

Do? Besides physics?

"Bah!" He threw up his hands. "I always thought this Eileen must be an interesting person—the only woman in the class, she asks such interesting questions. But now I find she is as boring as the rest! And you will only get more boring as time goes on. There's nothing you like to do but study physics?"

Well, I said, I used to enjoy playing tennis. I used to enjoy writing stories. But I figured I could wait to play tennis and write stories until I got out of grad school.

"Augh!" He covered his ears. "Do you know how often I've heard that? *I'll have time when I finish my dissertation.* Then there's postdoctoral work, establishing a reputation. *I'll have time when I get tenure.* Then there are labs to run, seminars and committees, politics in the department. *I'll have time when I retire.* By the time they retire, most of the men who've said that don't remember what it was they wanted to have time for. And I'll tell you, most of them come home at five thirty and get mad at their wives because the steak isn't on the table and they have to be back at the lab at six thirty. My wife and I don't have as much time together as I would wish, but at least I can recall what she looks like. We still enjoy each other's company."

This only confused me more. I was happy that my professor had a wife whose company he enjoyed. But I also felt jealous—I wanted to be that wife. If I became a physicist, who would be putting the steak on the table for whom? My professor was trying to provide me with a vision more humane than giving up everything I loved for physics. But now I needed to become a physicist who found time to play tennis and write books and put steaks on tables and spend time with a spouse whose company I enjoyed.

"I'm not saying any of this is easy," he continued. "My colleagues call me frivolous. Sometimes I go to the theater. Once, I took a few days' vacation after spending two weeks running experiments—you don't waste the time they give you on the accelerator, so you work all day and all night for two weeks. When I got my doctorate from

Columbia, I decided I wanted to do my postdoc work in Paris. My friends told me that I was crazy. They said that unless I kept my name in front of the research community, I would never get a position, that if I went to Paris I would disappear and everyone would forget me."

Stupidly, I asked if his friends had been right.

"Bah!" he said. "I'm at Yale, aren't I? And not all of them are here! I haven't won a Nobel, but I haven't done badly, either, and I doubt I would be any further along in my career if I had done my postdoc at Berkeley or Stanford."

After we bussed our trays, he said he would write me a recommendation only on the condition that I sign up for a class in creative writing. We took the elevator down, and as we stepped into the lobby, he said, "You will see. Someday you will write a novel, and you will dedicate it to me!"

For all you might think a liberal arts university would encourage its physics majors to take a writing course, fulfilling my professor's dictum turned out to be nearly impossible. Even today, the competition to gain admission to the few writing courses Yale offers can be intense. In the seventies, the English Department disdained such courses as neither rigorous nor intellectual. When I told an English major I knew that I wanted to take a writing course, he laughed and said I shouldn't bother—hundreds of English majors would be vying to get into the few seminars being offered, and I was better off signing up for a literature course. Disappointed, I asked which literature course I ought to take. Well, he said, I probably wouldn't get into Professor Bloom's seminar on Wallace Stevens, but there was no law against my trying.

And so I made the twenty-minute trek from Science Hill to Linsly-Chit, took a deep breath, and climbed the imposing marble stairs to the departmental offices, where I asked the secretary if I could please register for Professor Bloom's seminar on Wallace Stevens, not that I had any idea who Wallace Stevens was. If she had said, "Oh dear, I'm sorry, that's a very small class, and only the most dedicated English majors have any chance of getting in," I would have accepted my fate and found a less popular seminar to sign up for. But what the secretary did was laugh. "Are you kidding?" she said. "Do you really think you can just walk in here and sign up for Harold Bloom's seminar on Wallace Stevens?"

I might have been in eighth grade, listening to Mr. Van demand to know who I thought I was to take an advanced class in science. An English major asking to be enrolled in the third semester of Quantum Mechanics might have been met with the same incredulity. But how difficult would a physics major find even the most advanced class in English? When I tried to explain this, the secretary cut me off. "Look," she said, "if you want to sign up for a lower-level lit seminar, the list of sections that have openings is on the bulletin board downstairs." With that, she lowered her head and went back to typing.

Angry and discouraged, I inched my way down the stairs. But the scorn in the secretary's manner made me reluctant to consult the list of "lower-level" courses. Who was this Professor Bloom that I wasn't at a level to take his class? I was about to leave when I noticed a poster describing a nonfiction writing seminar. Because the course had only recently been opened, the instructor would be accepting applications until five the next day, each portfolio to consist of a short autobiographical statement and a sample of the student's writing. No mention was made of the applicant needing to be an English major, so I went back to my room and composed my autobiography, to which I appended the following short plea for acceptance:

> As a physics major, I have been rejected for all upper level writing courses at Yale. I have been laughed at for even applying. I do, however, intend to make writing my career—after graduate school in the sciences, after I've learned the subject about which I want to write. I am not interested in taking this course merely to be a dilettante, but rather to learn to be a nonfiction writer.

Was I professing this love of writing solely to gain admission to the course? I don't remember. Either way, I needed to figure out what sample of my writing I might submit. We weren't supposed to turn in an essay we had written for class, so my only choice was a profile of the mathematician and biologist Jacob Bronowski that I had published in the *Yale Scientific* the year before. Bronowski had died, and my friend Rochelle, who had become an editor of the *Scientific*, had asked me to write the profile, an assignment I agreed to undertake because Barry idolized Bronowski's work, especially *The Ascent of Man*, which he and I had watched on PBS. If you must be a scientist, Barry seemed to be saying, be a scientist who thinks and feels and writes. A scientist who is an artist.

Reading that profile now, I can see that two years at Yale had shaken my belief that Barry's ideal was possible. "Unlike the well-rounded Humanist of the Renaissance," I wrote, "modern researchers who refuse to specialize in one discipline usually fail to add any major systems of thought to science. If the men Bronowski so respected—the Newtons, Galileos, Bohrs, and Diracs—hadn't targeted their lives and work into narrow tunnels of science, he would have had nothing to extol in prose, nothing to teach laymen, and no pulpit such as television from which to lecture."

In fact, writing that profile nearly turned me off to writing anything else. In the late sixties, Bronowski had delivered the Silliman Lectures at Yale, and I interviewed a biology professor who so enjoyed Bronowski's presence on campus ("I had a dinner for him once and the evening fairly *crackled*") that he proposed to Kingman Brewster that Bronowski be invited to assume the mastership of a college. The appointment never materialized. As the professor put it, Brewster didn't think Bronowski "patrician enough" and looked down on the man for being so short and so like an actor. "Yale had the opportunity to have a true Renaissance Man." Here, the professor plucked his pipe from between his teeth and puffed a doughnutty smoke ring. "And Yale lost it."

I quoted him precisely, but an hour after I had dropped off the article for his approval (a novice mistake), he called and raged that he had never said any such thing. I know now he didn't want to go on record as accusing Brewster of not wanting to make a short, balding Jewish scientist with a heavy Polish accent master of a college. But at the time, I couldn't understand why a professor would lie about what he had said, and the experience soured me on writing—which made it doubly ironic that I should submit the same article to support my claim that I wanted to spend my life doing exactly that.

So many students applied for Martin Goldman's seminar in nonfiction writing that he narrowed the field to thirty and invited each of us for an interview. Yet another New York Jew, Goldman wasn't a threatening man, but I was nervous about meeting a senior editor of *Time* and sat in the bare stone cell he had been given as his office, awaiting what he had to say.

Apparently, he hadn't been impressed by my article about Bronowski. But he was intrigued that a physics major wanted to take a writing course, and he said something about C. P. Snow and how important it was to foster a dialogue between the "two cultures." I had never heard of C. P. Snow and couldn't have said anything particularly intelligent in reply, but when I again made the trek to Linsly-Chit to see if my name was on the list of students who had been admitted to the class, I was elated to find it was.

There are moments in a person's life when her perspective changes so radically she becomes dizzy with the dislocation. For years I had been sitting with my eyes closed, trying to imagine the farthest reaches of outer space, the neurons inside my brain, particles too tiny to be seen even with the most powerful microscope. Then I opened my eyes and looked around. Who were these people, and how they had come to be the way they were? Not: How had they evolved from protozoans? But: Who had they been as kids? What did it feel like to be them? For that matter, what did it feel like to be me?

I didn't know how to get any of this on paper. To demand your readers accept your assertions is like chalking equations on a board without offering the derivations to prove their truth. You need to re-create your experience in the faith this will lead your readers to experience what you experienced and draw the same conclusions. You need to search the data stored in your neural net and discover which details vibrate at the same resonant frequency as the other details connected to that experience, then release the hologram of that memory and bring it shimmering back to life.

A few weeks into term, our teacher asked us to take our readers on a tour of a place where we had undergone a powerful emotion. Rather than sum up the event from a safe intellectual distance, he advised us to daydream our way into reliving the adventure, then translate the dream to the page with as little thought as possible. This sounded like hippie hocus-pocus, but our instructor looked nothing like a hippie, so I decided to try his method, dreaming myself back to the boardwalk in Wildwood, New Jersey, and reliving the rare pleasure of feeling connected to my family, especially my brother, who gave up his usual teasing because he needed me to accompany him on the rides. On this particular night, we found ourselves in a car attached to the arm of a rickety, whiplashing spider and realized with dismay that I was so small I could slip beneath the bar and plunge to the pier below, at which point he threw his body across mine and clutched the bar, shouting that he would keep me from falling out and stunning me with the revelation that, despite the corrective punishments he usually meted out, he truly did love me.

When Mr. Goldman read my piece to the class, I received more recognition for my talents as a writer than I had earned in all the physics and math courses I had ever taken. The praise recognized not only my ability to complete the assignment, it commended me for who I was, in that only the person who had experienced what I had experienced could have written what I had written. Mr. Goldman read my pieces often. He never identified the essays as mine, but a classmate told me everyone knew because my leg started vibrating under the seminar table like a sewing machine. "You're really good," this classmate said, and the other students nodded. But rather than be afraid of their resentment, I sensed they had admitted me to their club. I listened to their essays and responded to what they wrote, as they listened to my essays and responded to what I had written. In any given hour, I learned more about my fellow writers than I had learned about my fellow physics majors in all the time we had shared on campus.

I began committing to paper every detail I could remember about elementary school, junior high, and high school, and my first two and a half years at Yale, culminating in a long final essay in which I recounted many of the incidents in this book. "With two semesters of college left," I wrote, "I often think I would like to drop physics and become a writer, but somehow, that seems like giving up. I have the feeling that the humanities are less serious, and therefore more appropriate for a woman, than science, and that people will say I ended up writing because physics was too hard. Even getting a PhD in physics with the intention of becoming a science writer seems like selling out all womanhood."

What I was struggling with was whether a scientist could be a complete human being. "While the eccentricity and intense narrowness of interest of a male science prodigy are not only excused but encouraged, girls are told to 'do other things' and be careful not to cut themselves off from a social life. A man who cannot take care of himself can find a wife who will, but there are few absent-minded women professors—if a woman forgets to wash her clothes or cook her meals, no one else will." The paper ends with this admission:

> I once agreed so much with a line from Yeats that I tacked it to my wall: "The intellect of man is forced to choose perfection of the life, or of the work." It was supposed to inspire me to devote myself to a life of science. To be a physicist, I thought I had to learn to live like a man. I no longer believe that to be true. Science does require a willingness to give up many aspects of a "normal" existence, and scientific progress demands a certain amount of specialization of interest. But scientists do not have to be martyrs. And the often artificially imposed formality of science, the mystique of the scientist fostered by an intentionally kept distance from laymen does not serve anyone's interest. As Jacob Bronowski wrote in *The Ascent of Man*, "Knowledge is not a loose-leaf notebook of facts. Above all it is a responsibility for the integrity of what we are."

Many of the students in that writing seminar were women, and I felt connected to them in ways I had never felt connected to any of my classmates in physics. Around the same time, I was joined in Quantum Mechanics by a senior computer science major named Leslie who had recently switched to physics. I was excited and afraid. What if she made a spectacle of herself, the way the ballet dancer had done in thermo? Worse, what if she upstaged me? What if she eroded my uncomfortable but special status as the only woman in the class? When I saw her that first day, I chose a seat as far from her as

I could. When she did nothing that might embarrass me, I began sitting closer, sometimes sneaking a look at her grades, wincing when she outscored me. When she sat at my table in the Kline lunchroom, I barely managed to say hello.

"Don't you think Professor Nemethy's great?" she gushed.

"Oh," I said, feeling even more threatened—Leslie had long straight black hair, a radiant smile, and a terrific figure. "Do you have a crush on him, too?"

"A crush? No. I mean he's a really good teacher. Why? Do you?"

Oh, no, I said sheepishly, all I meant was that I loved the course.

Leslie beamed. "Quantum's the best part of physics. I can't help studying it instead of working on my other courses. I got a job last summer at the Stanford Linear Accelerator. Particle physics is incredible."

"You worked at SLAC?" I said. "You must have been the only woman there."

She twirled her hair and laughed. "There *was* one other woman— she was running the group. Everyone was scared shitless of her, including me."

I asked if she hadn't felt out of place.

"Once in a while. I caught myself asking the guys for help with the electronics stuff a little too often—it seemed so natural, like calling a repairman to fix something when you know all about how it works and could fix it yourself. But other than that, I hardly noticed I was a girl. It's an advantage now anyhow. All the labs have to hire women because of affirmative action."

"Who wants to get a job because you're a woman?"

"Jobs are hard to get. I'll take one any way I can." She giggled. "Within reason."

"And that doesn't bother you? If I got into a place like Caltech or MIT because I was a woman, I would think I wasn't as good as everyone else there."

Leslie shrugged. "I know I'm as good as any of them."

But as the semester wore on, I became convinced Leslie doubted her abilities as much as I doubted mine. The only difference was that she had a sheer animal buoyancy that helped her to overcome those doubts. I tried to emulate her example. And where neither of us had ever had anyone to share our worries, we could now phone each other with desperate questions the night before a test.

One afternoon, we saw a notice that a physics seminar was being given by a postdoc named Beverly Berger, and we decided to go. With all the professors and graduate students in the department, we filed in to the cold, harsh lecture room in Kline, then listened as an awkward, poorly dressed young woman delivered a talk on gravitational theory. She was so nervous that when she wrote a formula on the overhead projector, we could see her hand shake. The moment she finished, several professors pounced to question her. I had never attended a postdoctoral talk and didn't realize this was standard procedure—all I could think was that if her talk had been a success, the audience would have applauded. When the first professor pointed out an inconsistency in her reasoning, I wound my fingers around the arm of my chair. When Beverly hesitated, I pictured her running out of the room in shame. The longer she stammered, the harder I squeezed the armrest. Finally, she drew some functions on the transparency and explained away the contradiction. Oh, sure, the professor said. I see.

Leslie passed me a note. "She shouldn't have shown him up in class like that. He'll never ask her out now." I wheeled around and saw Leslie grinning. "Let's go up to her afterward," I whispered, and Leslie nodded.

We waited until the professors had finished asking questions, then invited Beverly to join us for dinner. Sure, she said, then gathered her papers and followed us down the hill to Silliman. Beverly had gotten her degree before it became taboo for the men in her classes to make the sort of deprecating remarks the men in our classes had the decency to keep to themselves. But she assured us that she eventually had gotten used to the discomfort of being the only woman in the room. I asked if she was married.

"My husband teaches physics at Princeton," she said. "I drive down there on Friday nights, but we're both too tired to talk to each other—we eat dinner and zonk out by eight thirty. We have Saturday together, and I'm back here by Sunday noon to prepare my lectures and work on my research. It's not too great a situation, but there's no way out. I just try not to think about it."

Until then, Leslie and I had ignored what we would do about having families. But if what Beverly said was true, if you married a physicist—and who else were you supposed to marry?—both of you would need to get jobs in the same city. Even if you accomplished this feat, what would happen if you had a child? Your husband might consent to stay home and care for the baby while you were in the lab, and you could care for the baby while he was in the lab, but when would you see each other? I had never heard of day care—where I came from, mothers stayed home until their children were in school. What was I supposed to do, take six years off from physics to raise a child?

"I just want to make sure my kids can play tennis and ski," Leslie said, laughing, and we never discussed any of this again. The fault was mine, not hers. Here was the smart, funny, trustworthy female friend I had been looking for all my life (I was astonished when Leslie confided that she had been tapped for a secret society but refused to go along with the initiation when she learned she would be required to reveal a secret about someone she loved). But I found Leslie's constant smile and puppyish enthusiasm to be annoying. Or maybe it was just that Ed had let slip a sarcastic remark about how girlish and unserious she was. Worse, I was jealous. Even if I were to graduate a year early, the best I could do was to tie Leslie for the honor of being the first woman to earn a BS in physics from Yale.

Well, if I couldn't be the first, I would need to be the best. Thinking I would fulfill the requirement for an honors degree that I carry out original research by discovering some new particle, I asked Professor Zeller if he would take me on as his advisee. He didn't have any openings but suggested that I talk to Professor Bardeen, who was doing "a lot of cool theoretical stuff about black holes."

Black holes? Me? This was a time when few people who weren't physicists had even heard the term. Professor James Maxwell Bardeen, whose father had received not one but two Nobel Prizes, had studied for his doctorate under Feynman. He and a British physicist named Stephen Hawking had recently published a paper setting forth "The Four Laws of Black Hole Mechanics." Working on a research project about black holes with Professor Bardeen was as good as a free ticket to Caltech or Princeton, but I couldn't bring myself to approach his desk and instead hovered like some bit of celestial debris a foot inside his door. I stammered why I was there, thinking Professor Bardeen would say he had no project within the limited capabilities of an undergraduate. But he startled me by saying yes. Staring at his hands, he began to describe a series of calculations that might lead to the discovery of something called the Schwarzschild radius for a certain type of black hole. I was tempted to step closer so I could hear his quiet monotone, but the project seemed more appropriate for a graduate student who had taken three or four seminars in gravitational physics and relativity. Professor Bardeen seemed the center of a black hole himself. I excused myself and backed out, but he kept whispering to his hands without noticing I had gone.

That's when I remembered my idea of asking Professor Howe to be my mentor. He was a pure mathematician, which meant his work had absolutely no practical application. But in class, he would wander off from the derivation to meditate on hot new fields of applied math such as chaos theory or catastrophe theory. I was a little in love with him. And where I once would have considered him too brilliant to be approachable, I now had reason to suspect he was more human than he first appeared. My friend Fitz had been laid up with a high fever and swelling in his joints (apparently, he had picked up a mysterious new disease on an outing to Lyme, Connecticut), and Roger had showed up at the hospital to see how he was doing, although— Fitz couldn't help laughing—the guy had spent the entire time staring at his feet.

The first time I visited Roger in his office, I followed his suggestion that I remove the stacks of papers from the green Salvation Army couch; not knowing where to set them, I piled them atop the towers of textbooks on the floor. A paperback about fairytales lay spread-eagle across his blotter. His parka dangled from a bike twisted against one wall; above the bike hung posters of a Vietnamese liberation fighter, the Vogue Girl, the March Hare, a beached whale, and a cluster of finger paintings in runny crimson and muddy orange. When I asked if he would be my thesis advisor, he put his fist to his chin, looked up at the ceiling, and murmured, "A physics project? Hmm. I suppose it would need to have some sort of practical application." I lifted myself a few inches off the sofa to see him above the monolith of mimeographed articles on his desk. "I think you can take a group theoretical approach and prove that Huygens' Principle holds only in odd-dimensional spaces." When he saw my blank look, he lumbered to his feet and stepped over the piles of books on the floor to reach the blackboard. I needed five months to decipher the problem he chalked on the board that day—which only brought me to the point where I might begin to solve it—but I was fascinated enough that I agreed to take on the challenge.

I started by reading the work of the seventeenth-century astronomer and mathematician Christiaan Huygens, who believed light travels as a series of sharply defined spherical waves, this at a time when most scientists thought of light as a stream of particles. According to Huygens, if a candle were to flash, the radiation would spread like the skin of an expanding balloon; in a fraction of a second, this disturbance would reach an observer standing a few feet from the source, then travel outward, leaving no remnant of its passage—a good thing, given that electromagnetic signals would otherwise continuously assault our senses, one wave superimposing itself on the last forever. I had seen diagrams of a candle surrounded by concentric circles and taken for granted that light should behave this way. But the illustrators simplified their work by drawing the phenomenon in two dimensions. As it turned out, waves don't behave well in even-dimensional space. Rather than inflating like balloons, they expand like solid, swelling bowling balls (or, in the case of two-dimensional space, the cross-sections of solid, swelling bowling balls); even after the surface of the wave expands beyond a point, the disturbance keeps affecting that point forever. Because of friction, we don't usually notice. But if you were to toss a pebble in a pond of frictionless water, the circular wave would set a cork bobbing on the surface for eternity.

The question my advisor posed was why. Mathematicians had long ago predicted the phenomenon by traditional means. But Roger thought we could derive a more elegant proof based on the symmetries of the equations that govern the motion of waves in n-dimensional space, a method that would unite classical mechanics with quantum theory. Unfortunately, I knew nothing of this branch of math. I spent months trying to make sense of the books whose titles he had jotted on a pad. I would sit with my eyes closed, imagining how many times you needed to rotate a snowflake to get it where it started, or what a doughnut might look like in four dimensions, or how you might superimpose a set of gently curving sine waves to produce a jagged mountain range of a shape. I would trudge to Roger's office, wait for some equally frustrated graduate student to shuffle out, then clear yet another pile of papers from the couch, sit in the same indentation on the cushion, and peer over an even higher stack of papers on Roger's desk. Softly, slowly, he would get up from his chair, then pause, hunched beside the chalk tray, staring at the n-dimensional torus he was visualizing in his giant cupped hands, like a dreamy State Farm agent. I wanted so badly to see what he was seeing that I stared at that torus, too. But when I got back to my room, I no longer could see what I thought I had seen, and I remembered little of his explanation.

Laurel was still in Paris, and I grew more and more isolated, attempting to visualize what couldn't be visualized and becoming unsettled by an anxiety I had no idea was a reaction to the high dose of estrogen in the birth-control pills I had started to take at Ed's insistence. ("Sterile?" he said. "Are you kidding? There's no way I can sleep with you if you aren't on the Pill.") By now, Ed was stopping by my dorm room only for sex, but I wasn't sure I blamed him. I had trouble sitting still, and he must have wondered why I sneaked off to the bathroom after every meal. As I lay twitching in bed, a frenzy of images of multidimensional doughnuts alternated with a never-ending loop that showed me putting a pistol to my temple and pulling the trigger, accompanied by a soundtrack that chanted, "I'm going to shoot myself in the head, I'm going to shoot myself in the head." Exhausted, I hit upon the idea of keeping a bottle of wine by my pillow. One night, I awoke to find my roommate, Lee, hovering above my bed. "Lee!" I shouted. "What are you doing?" He apologized and confessed he had an interview with a top engineering firm the next day and doubted he could fall asleep without a slug from my Manischewitz.

As obsessed as I was about solving the problem my advisor had set me to solve, I remained determined to carry through on my promise that I take at least a few courses in the humanities before I narrowed my focus to physics. On top of Quantum Mechanics and Advanced Mathematical Methods in Physics, I signed up for a philosophy course taught by a young Pole who, as a prelude to studying Hume, had us sit around a seminar table trying to figure out how human beings would experience the world if we had only one sense organ: the nose. That left a single slot for my final course, which, if I wanted to graduate at the end of that semester, needed to be in physics.

As often happens, my failure to sign up for that one physics course was the result of powers I did and did not control. At the end of the previous semester, Martin Goldman had announced that John Hersey—the Pulitzer Prize–winning author and master of Pierson College—would be teaching an advanced seminar in nonfiction writing. As tempted as I was to apply, I still needed to take that last required physics course. And Hersey's seminars were even harder to get into than the seminar I had just completed.

What I didn't know was that Goldman had passed along my essays to Hersey and told him that I hadn't formally applied for his class because I was "too modest to blow my own horn." One afternoon, I came back to my room to find a note that Professor John Hersey had been trying to reach me. When I called back, Hersey chided me for mispronouncing his name—he had nothing to do with the chocolate company, he said—then told me that he was saving a place for me in his seminar, but dozens of other students were clamoring to take the seat and I needed to tell him right away if I was interested. I was mortified at having mispronounced his name. And I needed to take that last course in physics. But I stammered out yes, I would love to be in his seminar, at which he said I would need to stop by to get his signature on my permission slip.

Unlike the grubby, sterile offices on Science Hill, Hersey's office as master of Pierson College was a spacious, wood-paneled room with floor-to-ceiling bookshelves. I had never met anyone like him. Tall and long necked, he was as thin and straight as a totem pole, with a shock of white hair, impossibly elegant hands (he had given up the violin to become a writer), and searching, soulful eyes. He had a mellifluous voice, rich and deliberate; you could almost see him reach for the word he wanted, run it through his fingers and savor the way it felt before offering it in his palm to you. He was so . . . *patrician*, I thought. If this was the sort of man Kingman Brewster had in mind when selecting the master of a college, then no, short, plump, bald Jacob Bronowski wouldn't have made the cut.

If all this wasn't unnerving enough, when I reached in my purse for a pen so Hersey could sign my permission slip, I pulled out a tampon and handed it to the man. I seized the tampon back, but by that time I was so humiliated I excused myself and hurried out. How could I show up for his class? Then again, how could I *not* show up, even if it meant I would need to return to Yale the following year to fulfill my requirements for my degree in physics?

I was reluctant to tell my parents I wouldn't be saving them as much money as I had thought, but when my father found out I would be taking a course from John Hersey, he whistled and said, "No kidding!" Like the rest of the world, he knew Hersey as the author of *Hiroshima*. He also held Hersey in high esteem for *A Bell for Adano*, which portrayed the sort of petty military bureaucracy that so infuriated both men. And my father venerated Hersey—the son of Protestant missionaries—for having written *The Wall*, one of the first novels about the Holocaust and only the second book to win the newly created National Jewish Book Award. In fact, *Hiroshima, A Bell for Adano*, and *The Wall* constituted a very high percentage of the books my family owned.

The class met in a regally appointed seminar room in Pierson College, and when I walked in that first day, I saw seated around the table nine of the most attractive, well-dressed women and men at Yale, all of whom knew each other from the literature classes they had taken and the late nights they had spent putting out the *Yale Daily News*. Yet most of them were welcoming and kind. And, unlike my science classes, where I needed to leave who I was as a person outside the door, in my writing seminar, who I was as a person became the subject of my work. I handed in a group portrait of the secretaries and adjustors at the insurance agency where I worked and took my readers on a journey around my grandparents' Borscht Belt hotel.

In addition to critiquing our work in class, Hersey met with us privately to discuss our essays. I would arrive early, then sit on the stone wall that led to the Pierson courtyard trying to anticipate whether he would like what I had turned in that week. At the appointed hour, I would take my seat by his desk and watch him page through my essay searching for the tiny penciled ticks he used to guide him to the passages he wished to praise, or the weak, sloppy sentences he suggested that I improve. Then he carefully erased each mark, as if my essay were such a rare document that it might be desecrated by his penciled checks. Once, he asked what I thought of the criticism that a particularly pretentious classmate had leveled against my essay. Not wanting to appear defensive, I said I guessed she had had a point. Hersey stared out the window; the sun turned his fine white hair transparent. "Sometimes," he said, "when someone criticizes your work, you've got to tell yourself, *What other people say has nothing the fuck to do with what I know about my writing.*"

Considering how little we had in common—Hersey had never heard the term *Borscht Belt* and, if he had, wouldn't have been able to reconcile his romantic vision of the Jews who had perished in the Warsaw Ghetto with the guests at my family's hotel—I'm not sure why he took to me the way he did. Maybe he saw a bond between the shy scholarship student he had been at Yale and (as he put it in the letter of recommendation he wrote for me to get into the graduate writing program at Iowa) "the daughter of innkeepers in the Adirondacks who too often keeps her light hidden beneath a bushel." He encouraged me to apply to the fiction seminar he would be teaching the following year, and when a magazine editor in New York called to ask which students might contribute to an issue about "The New Youth," Hersey gave the man my name, with the result that an essay of mine was published that spring in *Life*.

That same term, I was asked to write an article for the *Yale Daily News* about why so few women majored in math. (The year before, only one woman had graduated with a degree in the subject.) Connie Gersick, the director of the newly formed Office on the Education of Women, confirmed what I had learned the hard way: the typical female freshman entered Yale trained only in algebra, "whereas the men are more likely to have had one or two years of calculus." Sheila Tobias had just published her groundbreaking article about women and math anxiety, and, disguising myself as the subject of an interview, I speculated as to why women might be more anxious about studying math than men.

Not only was I the only woman I interviewed for the article, as far as I could tell from the reaction the piece provoked, I was the only person who read it. And yet, it dawned on me that the obstacles I had been facing weren't mine alone and writing about such issues might prevent other girls from suffering what I was suffering. Soon after, Gersick persuaded me to help organize Yale's first conference on women in science, one of ten such conferences the National Science Foundation was sponsoring that year. As part of the day's events, I listened to Beatrice Tinsley—soon to be Yale's first female professor of astronomy—suggest that the cosmos, instead of collapsing into a fiery mass, might go on expanding forever, and I ate lunch with Margaret Mead, who delivered the keynote lecture. Sadly, I remember wishing that Margaret Mead weren't so frumpy (I had no idea she had been married three times and involved in two intense lesbian relationships, counting among her partners several of the most brilliant and passionate scientists of her generation) and that Professor Tinsley would do something about her hair (four years later, having only recently attained the resources and respect she so merited, Beatrice Tinsley would be dead of cancer).

For the last six weeks of term, I split my time between working on my senior project for Roger Howe and my final writing assignment for Hersey, which involved shadowing the director of neonatology at Yale-New Haven Hospital as he and his staff decided which of the drastically deformed newborns on their unit were beyond saving and eased them through euthanasia. Both projects were equally frustrating—the doctor said I could use his name, then reneged and said I couldn't—and equally exciting, leading me to suspect Hersey might be right that the solution to my dilemma lay in a career that involved both science and writing, even as I feared such a life would combine the least creative aspects of both endeavors.

Despite everything, I still planned to go on for my PhD, and when spring break rolled around, I convinced my parents to bankroll a flight to California so I could scout out graduate schools on the West Coast. Fitz's sister, who was only in high school but living on her own, said I could crash with her, and I spent my first days in San Francisco sighing over the sourdough bread, crab chowder, dark chocolate, and rich, strong coffee she set before me. I had arranged to meet with the director of graduate physics at Berkeley, and on the appointed day I showed up on campus. But as soon as I entered the physics building, I started to shake. With each laboratory I passed, with every problem set I saw tacked to a bulletin board, I grew sicker and sicker, until, by the time I reached the professor's office, I couldn't bring myself to knock. I continued down the corridor and walked out the other side, where I stood with my hands on my knees, gasping for air and trying not to pass out.

I had never not kept an appointment. Did this mean I couldn't apply to Berkeley? Even after I got back to Yale, I felt nauseated at the thought of applying to grad school. How could I keep working this hard for the rest of my life? How could I consign myself to spending decade after decade as the only female in the room? And yet, what would I do if I didn't do physics? More and more frequently, I found myself crouched in the corner of some room, gasping for air. The sicker I grew, the more drastically my relationship with Ed deteriorated. I was devastated when he broke up with me, but I didn't have time to brood: if I didn't think fast, I would end up spending yet another summer typing up accident reports and listening to the adjustors rate the titties of the women who danced at the Kit-Kat Lounge opposite the raceway in Monticello.

CHAPTER NINE

X-10, Y-12, K-25

Overcoming my parents' objections to living on my own required ingenuity. If I told them I would be working in a lab on campus, I would get the same argument I had gotten the summer before. So I came up with the idea of applying for internships at Bell Labs in New Jersey, the Fermi National Accelerator in Illinois, and Oak Ridge National Laboratory in Tennessee, the assumption being that my parents wouldn't be able to resist my working for the government or an industry as well known and reliable as the phone company.

The strategy worked. Professor Parker wrote me a recommendation for all three programs, and when I got accepted at Oak Ridge, my father barely murmured his objection. The only damper on my excitement was Mikis's remark that I had been offered the position because Oak Ridge needed to hire women. His theory didn't hold water—I had been rejected by Fermi and Bell Labs, which ought to have been equally anxious to fulfill their antidiscrimination quotas— but the comment ate away at my confidence, just as the hydrochloric acid had dissolved my stockings two years earlier.

Still, I couldn't help but feel special. Security clearance in hand, I flew to Knoxville and moved into the dorm on the University of Tennessee campus that had been reserved for us interns. The next day, we carpooled to Oak Ridge, where we followed a country road through the thick woods, with dirt paths snaking off beneath chained gates marked with fading Atomic Energy Commission logos. A tower seemed to step out from behind a copse of trees, and there it was, a Disneyland of accelerators, reactors, and smokestacks. WARNING, ROAD CLOSED TO PUBLIC, OFFICIAL USE ONLY read the sign, and even though my clearance wouldn't have gotten me into the Weapons Division cafeteria, I enjoyed flashing my ID to gain entrance to a facility the mere public couldn't enter.

That first day, each of us had our picture taken and clipped to a badge with radiation-detecting film. On our way back to Knoxville, we stopped for dinner. "You work at the labs?" The waitress pointed to my badge. "You're pretty young—you must be a really smart kid."

And I did feel smart. I had been hired because I had mentioned on my application that I knew FORTRAN, and even though I was afraid I would be sent packing as a fraud, it turned out I knew almost as much about programming as most of the employees in the Division of Neutron Physics, to which I had been assigned. My project was to design a simulation that could predict the pattern of radiation emitted from a nuclear reactor, and I soon became an expert on the shapes of smokestacks and the use of random-number generators to mimic the unpredictable path a radioactive particle might take as it was blown this way or that by the wind. My division head, a gruff, rotund, buzz-cutted Buddha named Tut, seemed impressed that I was majoring in physics at Yale, and he wasn't at all unsettled by my being female. His wife, Fran, struck me as dotty and dim, until I realized she was the real brains of the outfit, her brilliance dimmed only by the diabetic stupors into which she regularly fell. ("When she gets like that," Tut advised, "get crackers from the vending machine, stuff them in her mouth, and make her chew.")

Even a ten-week intern like me was awed by the Manhattan Project mystique that still surrounded Oak Ridge. The lab's symbol, imprinted on cafeteria trays and tourist handouts, was the Graphite Reactor, a fission pile built in 1943 as part of the crash effort to produce enough plutonium for the world's first atomic explosion. After the Army Corps of Engineers finished building the reactor, they erected the Y-12 National Security Complex down the road and the gaseous diffusion plant deep in the hills, both designed to enrich uranium for the Hiroshima and Nagasaki bombs. Those were the years of real secrecy—the older men liked to recall their wives' dismay upon arriving at their new barracks in the muddy Tennessee woods and their own inability to so much as hint at why they considered it a privilege to be working there.

At the start, I felt privileged to be working at Oak Ridge, too. I prided myself on knowing that the three sections of the lab—X-10, Y-12, K-25—took their names from their coordinates on a map. I loved hearing the old-timers tell stories about the famous physicists they had encountered in the lunch line. And I reveled in the camaraderie of my fellow interns, driving back and forth from Knoxville to Oak Ridge and taking weekend trips to explore the lush Tennessee countryside. I began dating a young physicist named Kevin, even as I reveled in the companionship of other women who studied science: Cheryl, who hated the fundamentalist college in Texas her parents had forced her to attend; Smitty, who spent the summer happily blowing up pipes in her ultrasecret lab; and Melinda, a Chinese American woman from California who attracted every guy in the group, including Kevin. I didn't even mind wheeling a shopping cart of punch cards across the parking lot every afternoon so I could run my program on the IBM mainframe, hoping I wouldn't get caught in the Tennessee downpours that thundered out of nowhere, compelling me to throw myself across the cart to protect my cards.

But I felt uncomfortable in the South. Kevin turned out to be a born-again Christian, as did many of the other students in our group. I overheard an intern from Louisiana call an intern from New Jersey a kike, and when one of the division heads invited us for a barbecue, I saw a billboard that proclaimed his neighborhood to be "restricted." Swimming laps in the UT pool, I noticed not a single black student. When the other women and I walked to town, men leaned from their cars and drawled: "Y'all wanna *fuuuuck*?" If we passed a fraternity, the boys on the roof held up numbers to rate our looks.

I also grew disillusioned with the us-versus-them mentality of the lab's employees. "We" were solving the energy problem; "they" didn't have a clue what a neutron cross-section was. "We" knew nuclear power plants were safe; "they" staged emotional protests, holding up the construction of new facilities. I grew tired of hearing the engineers ridicule the "morons" who thought it possible that a reactor might go critical (this, a year before Three Mile Island), even as they told stories about the janitor whose rubber boot had slipped off and gotten sucked into a pipe that fed a cooling tank, or the guys who had shielded a new reactor, only to run a Geiger counter along the road and find a spot so hot they scratched their heads in bewilderment—until someone pointed out they had neglected to shield the roof.

Our first week, we were required to attend a safety session where someone from Health Physics demonstrated how to detect radiation spills. As he passed his Geiger counter over the "suspected surface," the machine chattered a warning. But the demonstration was far from over. "Several radiation sources have purposely been hidden around this room," the man told us. "I want each of you to take a counter and see how many spills you can locate."

I said I would rather not play the game.

"Why?" he asked. "Are you afraid of the radiation? You don't honestly think we would expose you to harmful levels, do you?"

As a child, I had undergone a lot of unnecessary X-rays, and I didn't want to absorb even one millirem more than needed. But with everyone watching, I gave in and agreed to play the Huckle Buckle Beanstalk Radiation Game. I am sure the exercise added little to the level of radiation I received that summer. But I am equally sure it added little to my ability to detect a spill.

A similar episode occurred at summer's end. A dozen of us were asked to assemble at the base of two metal towers. "There used to be a reactor suspended up there," our guide said. The government had been trying to figure out how to shield a nuclear-powered jet. "Then Health Physics ran some tests and found out we were spraying radiation past the fence. That's why the reactor is underground. But man, you should have seen it hanging up there!" Shading my eyes, I looked up at the catwalk that spanned the towers and listened as our guide reminisced about the day a workman had been killed falling from the top. "But it's great up there on a clear day. You can all ride up to the catwalk for a view of the Melton Hill Dam."

One intern, Dave, glanced at the elevator crawling up one of the towers. "Any other reason we need to go up there?"

"If you'd rather not . . ." the guide said.

"Um, no, sure, I'll go up," Dave said, although he was clearly terrified of heights.

The only woman, I wasn't about to stay behind, and so I joined everyone else in the elevator, which lifted us to the catwalk. Two of the men immediately sat down and narrowed their vision to the tower, but the elevator was already dropping. I looked around, noting that yes, there was the dam. Then I glanced down through the open elevator-shaft and saw that the cage had stopped. "Uh, guys?" I said. "Something's wrong."

Sirens squealed. The yellow lights surrounding the towers started to flash. "Jesus," Dave whispered. "The reactor's going critical and we can't get down."

We were up there for an eternity, lights flashing, sirens blaring, until finally the elevator came back up and we hurried in. On the ground, we listened as the guide laughed and explained the prank. "Oh, we were just testing the radiation-detection system. Nobody's ever gotten that upset before."

As much as I enjoyed flashing my badge to gain entrance to the facility, once I was inside, I felt like an outsider. I hated the way the guys in mirrored shades who guarded the IBM mainframe made me beg if I wanted to learn the codes that granted me access to run my program. I resented getting laughed at for being a vegetarian, for not drinking beer, for spending my lunch hour reading *Moby-Dick*. I was appalled that the government had no way to judge the path of radiation being discharged by a nuclear power plant—my program wasn't bad at predicting the concentration of pollutants less than fifty kilometers from the source, but it broke down after that. And I was furious when Fran let slip that my work had been funded by the air force, which intended to use the simulation to predict the pattern of radiation emitted by a bomb so the next pilot could fly in with another bomb and avoid the radiation from the first.

The last night of the summer, Fran and Tut invited me to their house. Tut thanked me for putting up with Fran's diabetic stupors, and they asked me to consider coming back to work for them after I finished my degree.

How could I? I said bitterly. Why hadn't they told me that my project was funded by the air force?

Fran blinked. Who did I think ran Oak Ridge? Tut asked. Hadn't the lab been started to develop the atomic bomb?

Now I was the one who blinked. Hadn't I read *Hiroshima*? Part of the atomic bomb described in Hersey's book had been developed right here at Oak Ridge, and the other part at Los Alamos by Richard Feynman. Softening, I hugged Fran and Tut good-bye and said I wished I could come back to work for them, but I was determined to study theory, although even as I said this, it came to me that theoretical physics had been just as responsible for the bomb as practical research like Fran's and Tut's.

When I got back to New Haven, I wrote a column for the *Yale Scientific* expressing my concern that the us-versus-them mentality at the lab served to insulate Oak Ridge from the public's legitimate fears about nuclear power. Almost immediately, I received a call from an official who snapped something about my biting the hand that fed me and reminded me that I had agreed not to publish anything about my research without clearing the article through proper channels. I hung up, then sat shaking, queasy with the notion that I was a traitor to my government and would never again receive a security clearance to work for my country—an especially upsetting notion for someone contemplating a career in physics. Finally, I convinced myself I didn't care. What scientist worth her salt would want to work for an institution that would blackball her for writing the truth as she had observed it with her own two eyes?

CHAPTER TEN

Life on Other Planets

A summer spent plotting the path of a plume of radioactive vapor convinced me I would go on in physics only if I were talented enough to study completely useless questions about space and time. As soon as I got back to Yale, I signed up for a graduate-level seminar in relativistic astrophysics, which promised to cover the topics I had loved reading about as a kid. If I had been excited buying my giant blue copy of *Physics* freshman year, I approached my gargantuan black copy of *Gravitation* with the same awe that inspires the monkey-men in *2001: A Space Odyssey* to raise their hairy hands to the big black monolith.

The textbook had been written by the three fathers of the new cosmology, Charles Misner, Kip Thorne, and John Archibald Wheeler, and they began their tome with a solemn chant that reads like the initiation for a government-supported priesthood:

We dedicate this book
To our fellow citizens
Who, for love of truth,
Take from their own wants
By taxes and gifts

And now and then send forth
One of themselves
As dedicated servant,
To forward the search
Into the mysteries and marvelous simplicities
Of this strange and beautiful Universe,
Our home.

Every member of the priesthood whose image appeared in the book, from Galileo to Robert H. Dicke, was male, but I wanted to join the order. I wanted to be initiated into "The Parable of the Apple," "The Mathematics of Curved Spacetime," "The Evolution of the Universe into Its Present State," and, with its title right out of *Star Trek*, "Frontiers," which promised to reveal what happens "beyond the end of time" when our universe will be "reprocessed."

Today, when I open any of the other books I used as an undergraduate, the mold-spotted covers release the memory of the person I was when I took the course. Leafing through *Gravitation*, I can't understand a single page, and no memories of the class waft out with the mildewed smell. Of all my classes, Relativistic Astrophysics is the only one whose instructor I can't recall—he was no one I had a crush on, neither terrible nor outstanding. It was simply a graduate course in the field that interested me the most, and a course I had absolutely no problem acing. I was the only undergraduate in the class, but when I look now at my final exam, I see nineteen pages of equations scribbled in handwriting that must be mine, and a bold red 96, which, while unaccompanied by a single comment ("Good job!" or "Nice work!"), allowed me to earn an A.

That semester, the physicist Sheldon Glashow visited Yale to give a lecture. Glashow, already a celebrity for predicting the existence of the charm quark, would soon be awarded a Nobel Prize. Whether because I was doing well in the class, or because our visitor wanted a young woman to enliven the company, my professor invited me to accompany Glashow to lunch. Afterward, I called my father to crow. "Just remember that even famous physicists put on their undershorts one leg at a time," he said. I wanted to strangle him for deflating my accomplishment. But for the rest of my life, whenever I felt intimidated by a famous man, I imagined him putting on his undershorts, one leg at a time, and my timidity disappeared.

To graduate, each physics major was required to turn in a paper surveying the articles in a field. Most of my classmates chose small, clearly defined topics in particle physics or materials science. But I wanted to satisfy my curiosity as to what, if anything, my elders were doing to contact life on other planets. The reading was easy—relatively few formulas are required to determine the probability of extraterrestrial intelligence—and I was heartened to learn that scientists at MIT, NASA, and Hewlett-Packard took the search for extraterrestrial life as seriously as I did. That I might someday be part of the team that communicated with beings on other worlds was so thrilling I decided to risk my professors' scorn by turning in a paper in which I supported such endeavors.

Unfortunately, I had nothing to show for all the months I had been trying to understand why waves in odd dimensions expand as hollow balloons while waves in even dimensions propagate like solid bowling balls. As the leaves in the Silliman courtyard turned yellow and drifted to the ground, so did the wads of crumpled legal paper accumulate around my desk in the miniscule room I had acquired in that year's lottery, having used up my senior standing the year before. And yet, I hung on to my sanity, in part because I had signed up for a seminar called Philosophy in Literature taught by Maurice Natanson, the professor with the Talmudic beard who had tutored me in the fine art of existing. I was drawn by that preposition in the title. How *did* philosophy get into literature? Did the author fashion his novel the way a child sculpts a pot, then pour in a few quotations from his favorite philosopher? Or did he read the philosophy book first, then write a novel to illustrate what he had read?

The first day of class, I raised my hand and asked. The student beside me sniggered, but Professor Natanson stroked his beard and said, "That's an extremely perceptive question." Certainly, a writer might be influenced by a philosopher. But for the most part, it was simply the case that great writers found themselves preoccupied by the same questions that preoccupied great philosophers. Neither the philosophers nor the writers could ever truly answer the questions that preoccupied them. But the writer and the philosopher shared the same concern for the deepest, broadest areas of human existence.

Oh, come on, the student beside me sneered. That's like saying people who take their philosophy from song lyrics are as profound as people who read Wittgenstein.

Professor Natanson raised his magnificent eyebrows. Ah, he said, but who is to judge that songwriters aren't as profound as novelists and philosophers? All three are concerned with what it means to fall in love, to work, to be a parent or a child, to face the fact of one's mortality.

The student didn't appear convinced. But that answer gave literature a seriousness I hadn't been willing to grant it. Unlike a physicist, a writer could be concerned with anything and everything, including what it meant to be a creature that could fathom the complexity of the universe, as well as the hovering shadow of death that would end her consciousness. The novels we read moved me and made me think like nothing I had read before. How could I not appreciate *The Magic Mountain,* in which a mediocre engineering student learns that the true subject of science, of all study, is "the human being, the delicate child of life, man, his state and standing in the universe"?

But the assignment that moved me the most was "Bartleby, the Scrivener" by Herman Melville. Bartleby, the human copying machine, decides he *prefers not to* continue copying the legal papers his employer wishes him to copy. I poured a heart I didn't know I had into writing that essay, and when I got it back, I saw that my professor had responded not with a grade, but with a note saying he wished to become better acquainted with the student who had written such a paper.

Maurice Natanson was the rabbi you wish had taught Hebrew school when you were a kid, although he was a rabbi who didn't appear to believe in God. Not only had he heard of the Borscht Belt, he had gone to camp there. He was a passionate chess player, and this particular camp had advertised chess as one of its specialties. But when my professor—whom I pictured as having a flowing white beard even when he was a child—sat down to challenge the camp's so-called master, he beat the impostor soundly.

Here was someone who knew where I was from and understood how difficult it might be for a student who had just spent three and a half years proving she could excel in physics to walk away with no better explanation than she *preferred not to.* "Having read your essays this term," my professor said, "I can well imagine that you have the talent and sensibility to become a writer."

In my three and a half years at Yale, not a single physics or math professor had hinted I might be gifted enough to achieve my dream to become a theoretical physicist, while five of my instructors in the humanities had come straight out and said I had the talent to become a writer.

It's impossible to overstate how lonely I felt as the only woman in my physics and math classes compared to the camaraderie and encouragement I felt in my classes in writing or philosophy. Leslie had graduated. Laurel was busy applying to medical school. Kevin, my boyfriend from Oak Ridge, lived hundreds of miles away, and both of us were too poor to visit. My only real human contact came when I worked as a dishwasher in the dining hall. Working in the dining hall not only paid well, it got me out of my room and away from the legal pads on which I kept scribbling the same Fourier transforms and Bessel functions. I felt more comfortable chopping onions in the kitchen than anywhere else on campus; the yelling, sweating cooks, the bubbling pots, the stainless steel tables, the giant cans and can openers reminded me of Pollack's Hotel. I knew how not to antagonize a chef ("You leave my favorite knife on metal table, I kill you!"). I enjoyed joking with the Irish and Italian women who staffed the kitchen—they reminded me of Rose, Anne, and Gladyce from the insurance company back home—and I felt more relaxed shucking corn with my fellow work-study students than I did attending my seminar in gravitation with my fellow physics majors.

The only part of the job I disliked was ladling sole Florentine to classmates who stared past me as if my blue work-shirt and paper cap rendered me invisible. I asked to be reassigned to the dish room, where I exulted in my ability to heft a rack of tomato-encrusted soup bowls onto the conveyor belt, then swing that same rack, the bowls now shiny clean and steaming, onto a cart and ferry it to the serving line. The other dishwasher, Willie, a goateed songwriter from New Haven, would croon his compositions in my ear and—when the garbage disposal grew clogged with napkins and then erupted, spewing me with chopped meat and spaghetti sauce—cajoled me out of my foul mood by singing, "Isn't this a beautiful day? It's the Lord's blessing. I just found out I have a singing engagement for the weekend in New York. The Lord bestows His blessings on us all."

I wasn't sure how seriously to take Willie's claims about those singing engagements—until he came in one afternoon and crooned that he was quitting because he was now able to support himself through his music. "Just keep believing in yourself, sister. The Lord will bestow His blessings on you, and you will be able to get out of here and make your dreams come true, like I did."

I was such a baffling mix of confidence and insecurity, I felt as if I truly were a dishwasher who only happened to be attending a graduate class in gravitational physics and a seminar called Philosophy in Literature taught by one of the country's best minds (before he came to Yale, Maurice Natanson had been Donald Barthelme's teacher at the University of Houston, a relationship Barthelme credited for his development as a writer). In the journal I kept that year, one entry reads: "There is the me that thinks she is nearly perfect, and the me that thinks she is shit, the me that knows how to be happy, and the me that refuses to want to be happy at all."

I despaired of being smart or creative enough to solve my project for Roger Howe. But deadlines have a way of compensating for a lack of inspiration, and in the one week before my project's due date, I gained nine months' worth of understanding. In a daze, I pushed myself on, scrawling pages of derivations interspersed with "So we see" and "Thus we know," until the problem I was solving, which constituted no more than a corpuscle in the body of scientific research, loomed in my mind as the greatest discovery in mathematical physics since Newton invented calculus. My paper was due that Monday. On Friday, I was working in odd-dimensional space when integrals and omegas started dropping out of my calculations, and the equations simplified to Dirac deltas—spikes of infinite height and infinitesimal width, each centered on a single point.

Dirac deltas! How cool was that? In the crowded zoo of mathematical creatures, the Dirac delta plays this very useful role: the delta picks out the value of a function at only one point and wipes it to zero everywhere else. In three-dimensional space, the mathematical machine that propagates a wave turns out to be a combination of Dirac deltas that allow the wave to live only on the surface of a spherical shell. When I switched to even-dimensional space, the deltas disappeared and the propagator became an ugly but benign monster that allowed the wave to live not only on the surface of the sphere but everywhere inside. Why this was true—why three-dimensional space supports human life in such a way that waves of sound and light affect us only for an instant and then quickly die out, while our existence in two or four dimensions would be a hellish cacophony of noise and electromagnetic radiation—still lay beyond my ability to comprehend. But right then, with all those magically spiky deltas popping up like unicorns on the final legal pad in my pack, I wanted nothing more than to spend my life thinking about such beautiful and perplexing questions.

After nine months of missing parties, cutting dinners, and losing sleep, I had at last solved the problem my advisor had challenged me to solve. Maybe my satisfaction in glimpsing those beautiful Dirac deltas should have been its own reward. But I wanted desperately for my advisor to acknowledge what I had accomplished. He must have been pleased. Yet I don't remember him praising me in any way. Then again, he had known the answer before I started; if not, he could have figured it out in a few hours. I was dying to ask if my ability to solve the problem meant I was good enough to make it as a theoretical physicist. But that would have been like asking a lover if I was pretty; if you needed to ask, you weren't.

Besides, I was afraid to steer our conversation beyond the boundaries prescribed by the propagation of n-dimensional waves. I hadn't forgotten the time I had been sitting at my professor's desk, watching him work some proof, when his office door blew shut; instantly, he jumped up and opened it. Not that I could blame him. I had crushes on them all—Roger Howe, Mike Zeller, and Peter Nemethy. My attraction to my professors kept me working to please them long after I might otherwise have given up. Yet that same attraction made me too self-conscious to ask them for guidance and, in some cases, may have made them too wary to provide it.

Everything about my time at Yale was a contradiction. The *Yale Scientific* ran articles about the research being carried out by the bespectacled male engineers in the accompanying photos, along with ads inviting adventurous Yale grads to enroll in the navy's ROTC program and operate nuclear-powered subs. But I'm not sure what the magazine's readers made of a column called "Notes from a Black Hole," in which a female undergraduate attempted to bring the principles of the New Journalism to bear on her experiences on Science Hill.

My first essay, "Confessions of a Faltering Physics Student," was written in the form of a journal kept by a fictitious young man named Kevin. Recognized by his high school teachers as exceptionally gifted and encouraged to skip ahead two years in science and math, "Kevin" enters Yale as a cocky freshman, only to earn a 32 on his first exam. He buckles down and pulls all As, but Kevin worries that he earned his grades through brute force rather than creativity. He considers switching to the humanities but can't find the courage to tell his friends that he is giving up physics for English. ("That's what physics majors say as a joke when they're walking out of a room after a hard exam.")

Rather than enroll in a seminar on James Joyce, he forces himself to sign up for Quantum Mechanics, where his passion for science is rekindled by the crazy notion that "the world is composed of red, white, and blue quarks tied together by strings or contained in cellophane baggies." If you study the subject long enough, Kevin confides to his journal, "quantum becomes your secret, a fantasy version of the universe. You're tempted to tell the people standing ahead of you at the cash register that they are wave packets, or that particles can tunnel through infinite potential wells, but you decide to keep your knowledge to yourself. They wouldn't understand anyway. And you begin to hope you will be able to add a new sentence or two to the fairy tale."

Accepted to Berkeley, "Kevin" moves to California, where he vows to give up "Aristophanes, Whitman, Joyce, and Updike," "live a tuna fish-can-to-mouth existence without complaint," and "vow celibacy until such time (if ever) that I can maintain a loving, two-sided relationship with a woman. I will not marry until I can remain at the dinner table for at least fifteen minutes after finishing my meal before having to rush back to the lab to watch a crystal grow."

I wrote that first column from the point of view of a white Christian male because I wanted my readers to see themselves in what I wrote. (The first time I ever went on the web, I noted with pride that someone at MIT had posted the column on his board.) By the second installment, "Inside the Oak Ridge Fence," I was writing in my own voice. And by the last, I had found the courage to describe what it felt like to be a woman majoring in physics at Yale, starting with the Catastrophe of the Smoking Stockings and ending with the barbecue at the home of my division head at Oak Ridge, where I had felt sorry for the man's wife and retired with her to the kitchen while the men continued their discussion in the backyard, the male interns acquiring a list of researchers they could contact when they applied to grad school.

"These incidents may seem trivial," I wrote. "They're certainly not the stuff of which lawsuits are made. I think we've eliminated most of the concrete barriers that have kept women out of physics in the past—the obstacles I faced were mainly psychological. But many women must still be prevented from giving in to unfounded anxieties about their ability to study science." That I was one of these women the essay neglects to mention. Instead, I offer the optimistic conclusion that the "biggest reason there are so few women in the physical sciences is that there are so few women in the physical sciences," and the prediction that the problem is nearly solved because "two women will graduate from Yale with degrees in physics this May—a 100 percent increase over last year's total."

As with everything I did, the consequences of my writing that column were paradoxical. One male reader admitted he had been so lost in his physics classes he didn't understand what he didn't understand, but like the other boys, he had been too proud to let on. Besides, he said, they knew they could count on me to ask. Professor Zeller stopped me on the street to say I had changed the way he and his colleagues viewed not only women in the sciences, but their own education. And yet, only one female student commented on the column: at breakfast, Erika informed me that I hadn't expressed a single thing she felt about majoring in physics. Stunned and hurt, I decided I would never again presume to speak for anyone but myself. But if I could speak only for myself, what was the point in writing?

Still, I took seriously Hersey's suggestion that I combine my two passions and write about science. A pair of mathematicians had been invited to discuss their solution to something called the four-colour map problem. For centuries, cartographers had suspected no more than four colors were required to differentiate the regions on a map, but no one had been able to prove this conjecture. Two professors from Illinois had programmed a computer to test every possible counterexample of the rule; when the computer failed to find a single map that required more than four colors, they pronounced the theorem proved. Most mathematicians cried foul. Proofs were supposed to be "beautiful" or "elegant," a series of logical statements that followed one from the next, step by deductive step. Listening to such objections, I wondered why it wasn't beautiful that human beings had invented a machine as elegant as the digital computer and derived the software to prove a theorem that couldn't be proved any other way.

Thrilled by the first original idea I'd ever had, I spent weeks researching the history of the mathematical proof and then wrote a twenty-page article for the *Scientific* arguing that computer-based proofs met a new definition of beauty. My editor was impressed but said she needed to give my essay to a professor in the Philosophy Department to see if my arguments made sense. When she handed back my essay, I was sickened to see a dozen red-inked queries in the margins. For all I know, the questions were designed to help me strengthen my arguments. But I couldn't make out the professor's writing and assumed if he thought my paper had merit, he would have added a note that said so.

I have no way of evaluating that paper now: I was so ashamed I tossed my only copy in the trash. For decades, I could barely look at a map without wincing in humiliation. Then, a few years ago, I was asked to participate in a forum about a novel based on the life of the Indian genius Ramanujan. A member of the audience asked a mathematician on the panel what he meant by saying a good proof needed to be "elegant." The mathematician offered the standard definition, then cited the computer proof of the four-colour map problem as an ugly, brute-force solution. I countered with an abbreviated version of the paper I had written decades earlier.

"Why, I never thought of that!" my fellow panelist exclaimed.

A wizened old man in the audience identified himself as a former chair of the Math Department. "Your comment strikes me as profound. Perhaps, when we say that a proof is elegant, we are not as certain as we think that we know what we mean."

That Christmas, after I had finished the coursework for my degree, I went home for a depressing winter break. I had put on weight, and my family didn't hesitate to inform me that I looked like a balloon. Then I returned to New Haven and moved into a roach-infested apartment a few blocks beyond Science Hill. Walking home from the dining hall each night, rank with the effluvia of the dish machine, I felt as if I had never gone to Yale, as if I were too poor and graceless even to apply.

All that saved me was Hersey's decision to let me audit his fiction-writing seminar, even though he would need to critique an extra student's stories. At first, I was alarmed by how few of the authors Hersey mentioned in class I had even heard of. Thankfully, it wasn't too late to acquire the background I had missed. I sat in my roachy apartment reading every author Hersey mentioned, as well as *The Second Sex* by Simone de Beauvoir and the diaries of Virginia Woolf. But I might never have become a writer if Hersey hadn't passed out copies of a story called "Goodbye and Good Luck" by someone named Grace Paley. Until then, I had never heard an author's voice that reminded me of my own. By handing out Paley's story, Hersey provided me with a way to see and hear myself as a writer, as I could not see or hear myself as a physicist.

The week our first stories were due, I read Paley's *Enormous Changes at the Last Minute*, then took a running leap into my own story, which I started and finished in one heart-racing afternoon, and even though I wrote from the perspective of a man—a mime, of all things—I wrote in a voice that would have been my own, if I had been a male mime. I wrote about people I knew—my parents; and I wrote about a place I knew and loved—the Jersey shore. It was a beginner's story, but I used it to express all my contradictory feelings about my parents, how ashamed I was of their vulgarity, how much I loved them and wanted their approval. I was stunned when the phone rang and a deep, deliberate voice said, "This is John Hersey. I have just read your story, and I wanted to make sure you knew what you had done. I wanted you to understand how much you moved me."

Me? I had moved John Hersey? "Thank you," I said, then hung up and wept.

In truth, I was as much in love with our writing professor as I had been crushed out on Michael Zeller and Peter Nemethy, the difference being that Hersey was too old for me to consider him a romantic partner. What he provided was a vision of the sort of man I had never known existed, a man consumed in the pursuit of beauty. The beauty lay not in the equations that governed the workings of subatomic particles, but in the sentences that described an encounter between a father and his child, in a gesture or a bit of dialogue. Once, when I told Hersey I heard the rhythm of the sentences I wanted to write before I knew what those sentence might contain, he nodded with the delight of a man who regularly listens to the music of the universe and has encountered a young acolyte who can hear and hum the same harmonics. In class, he would allow us to blather on about a story, then make a few eloquent comments that revealed what was emotionally true or resonant. *Resonant*—I remember that word, because I was struck that a line in a story might vibrate at exactly the right emotional pitch to strike a sympathetic cord in a reader's chest. As a writer, Hersey said, your job wasn't to pass judgment on your characters; your job was to understand people unlike yourself, to inhabit their souls. He gave a name to the state of being I had been trying to define since childhood, a state he defined as *grace*, not in the religious Catholic sense, but in the sense of transcending the usual limitations on your capacity for empathy or compassion. Even as he gave me stories to read that taught me to love the crazy Yiddish roller coaster of my own voice, he taught me to love the more dignified possibilities of the language as spoken by the son of a Protestant missionary. Yet he was anything but a pedant or a prude. When one of my classmates informed us that he had agreed, on the advice of his guru, to give up sexual intercourse, the pain was evident on Hersey's face. Why, he said, would anyone agree to relinquish the most pleasurable human activity in this harsh and lonely life?

One of the best aspects of Hersey's class was that half the students were women. In physics, no matter what honor I received, I could hear Mikis's voice assure me the honor had come only because I was female. When I found a letter in my mailbox from John Archibald Wheeler—the same John Archibald Wheeler who had written *Gravitation*—and learned I had been nominated to participate in a conference in theoretical physics to be held that March in Austin, and would I please submit a description of the research I intended to present if I were invited, I brushed off the honor on the grounds that Professor Parker had nominated me because he knew Wheeler was desperate to find any female undergraduate doing research in theoretical physics.

Only when I received the notice that my proposal had been accepted ("Happy we are that you will be joining us for the Second InterAmerican Undergraduate Conference in Theoretical Physics," the Yoda-like missive read) did I allow myself to believe that John Archibald Wheeler wouldn't be flying me to Texas and putting me up at a fancy hotel out of obligation. He wouldn't be sending me instructions about how to present my research ("Often talks are made measurably easier to follow if the key ideas are put in really big letters on transparencies") if he wasn't interested in hearing what I had to say. And he wouldn't be seating me at a table with Steven Weinberg and Ilya Prigogine if he didn't assume I would be able to keep up my end of the conversation.

By the second day of the conference, I realized that the other two women who had presented their work seemed as qualified as the men. The members of the Physics Department took us to their houses for "fireside chats" about the advantages of studying at UT Austin, then drove us far into the Texas hill country to dine at a barbecue joint only the locals knew. The countryside was spectacular, and after an hour's drive I was famished. But the only vegetarian items on the menu were coleslaw and corn bread, and as the men chewed through platters of ribs and downed pints of beer, I felt more and more removed from their company. The boys were a weedy, awkward lot. (When did male physicists start to look like Michael Zeller?) Then again, when Wheeler sent each of us a photo of the participants at the conference, I was horrified to see that I had put on so much weight my breasts tugged against my shirt, my cheeks had gone round and pouchy, and the big plastic glasses I had taken to wearing caused me to resemble Rocky the Flying Squirrel.[1]

After I returned from Austin, I walked around Science Hill trembling on the brink of applying to graduate school. If those seventeen boys and two girls were my primary competition for success in theoretical physics, why shouldn't I get into Princeton? I began to look forward to the night we physics majors would be presenting our honors projects to the department. Having delivered exactly such a talk in

---

1. Along with the photo, the conference organizers sent us a list of the participants' addresses, urging us to stay in touch. Googling each name now, I see that all but one of the seventeen men became physicists, although most became experimentalists, and one gave up his career as a theoretician to price exotic options for Morgan Stanley. Of the three women, one earned a master's degree in math from Princeton and now runs "creative math camps"; one became a writer; and only the third acquired a PhD, achieving a distinguished career in chemistry and astronomy.

Texas, I was spectacularly well prepared, with a memorized speech and a series of transparencies inked in really big letters. Roger predicted that the faculty would ask if my work had any practical applications, and he suggested I do a bit of research about the laying of the first transatlantic cable, which had failed to produce an audible signal because the engineers neglected to take into account the cable's thickness (in two dimensions, the sound waves muddied each other's signals rather than traveling in sharp, clear fronts). Roger didn't remember where he had read this information, and the only reference I could find was a vague mention on the first page of my freshman textbook. But surely the faculty would be impressed that I was the only member of the class who had done a project in theoretical physics, which everyone agreed was harder than applied.

Nonetheless, when I slid my first transparency onto the projector, the department chair, an august, white-maned personage in a striped red vest and bowtie, closed his eyes and started snoring. Only when I finished speaking did he jerk awake. "I don't suppose—" He lifted one thick hand lackadaisically in the air. "I don't suppose your research has the slightest practical application. Would you say?"

"Why, yes!" I said. "As it happens, when the engineers who designed the first transatlantic cable—"

"Very good, very good." He put his hands to his knees, pushed his bulky, august personage from his desk, and left.

A few weeks later, I had the even more disheartening experience of receiving the results of my GREs. I had earned a high score on the verbal section, but my math score made me cringe. I had been suffering from a cold. And I hadn't bothered to study for an exam designed to test how well the average humanities major could add, read graphs, and calculate percentages. But if I had done so poorly on a math exam that was meant to be taken by comp lit majors, how much lower would I score on the graduate exam in physics?

The following week, Professor Parker held a meeting about how to apply to grad school. After listing the top few institutions under the letter *A* and the next tier under *B*, he described the trajectory of a typical career in experimental physics—the years devoted to completing a postdoctoral fellowship, the cutthroat competition for an academic job, the struggle to win grants and publish articles.

Becoming a theoretician was so much more daunting that he didn't mention the possibility; if succeeding as an experimentalist was like climbing K2, then gaining tenure as a theoretician was like soaring to Neptune powered by nothing but the nuclear-powered reactor inside your head.

As usual, I was the only student who asked a question. "What happens to people who get degrees from those B-rated schools?"

Professor Parker rubbed his chin. "I guess they go into industry," which caused everyone in the room to laugh, the way they might have laughed if someone asked, "What happens if you can't get a really beautiful woman to marry you?" and the answer had come back, "You marry an ugly broad and close your eyes when you're having sex."

No, no, Professor Parker said. A lot of really cool work was being done in industry. Nearly everything in solid state, thin films, and lasers was being funded by corporations or by the government.

The government, I thought. Corporations. If you went to a B-rated grad school, you ended up designing smokestacks for nuclear power plants or figuring out when it might be safe to send in another pilot to drop a bomb.

What upset me even more was that my advisor hadn't said, "But you *aren't* mediocre, Eileen." He might not have wanted to say that in front of the class. But couldn't he have taken me aside and assured me that if I didn't want to work for a corporation, I needn't work for a corporation? He might have said that even if my GREs turned out to be a bit low, few other candidates would be able to boast the sort of research I could list on my application—the scattering program I had written for Michael Zeller, the diffusion simulation I had developed for Fran and Tut, the problem in advanced mathematical physics I had solved for Roger Howe. Why didn't anyone ask what graduate schools I was applying to? Why, when I mentioned shyly to Professor Zeller that I hoped to get into Princeton, did he shake his head and say that if you went to Princeton, you had better put your ego in your back pocket because those guys were so brilliant and competitive you would get that ego crushed? Even though I had earned all As in the most advanced undergraduate theoretical physics courses offered in my department and in a graduate course in gravitation, his remark made me feel as if I weren't brilliant or competitive enough even to apply.

Who knows, maybe if I had gone to Princeton, my ego would have gotten crushed. Or maybe—and this might have been my real fear—I would have discovered that I was bright and competitive enough to keep up with the pack, which meant I needed to spend the rest of my life snarling and snapping at the heels of a bunch of sexist assholes like Mikis, making myself so mentally ill I couldn't walk down a flight of stairs without imagining my skull cracking open, and continuing to wear baggy shirts and unflattering hairstyles and the biggest, ugliest glasses I could find so I might be taken seriously as a scientist.

Besides, who was I kidding that I could keep up with my writing while I was earning a PhD? It was easy to imagine being a writer who read books about the latest advances in particle physics and cosmology, but not a theoretical physicist who found the time to read stories by Grace Paley, let alone write stories of her own. What if I spent five or ten years studying for my PhD, only to discover I wasn't nearly as creative in physics as I was as a writer? What if I needed to spend my life depending on men to assign me projects and tell me if my work was good or bad? I wish I could say I made a rational decision to give up physics. But mostly what happened was that I found myself sitting in John Hersey's office one afternoon, and I looked across the desk, and I thought: *I want to spend my life the way you're spending your life. I want to keep writing stories. I want to keep reading books. I want to send you a novel I've written and hear you say it moved you.* And the next thing I knew, I blurted that I had decided to become a writer.

I was afraid Hersey might try to talk me into staying in physics. He had warned us a writer's calling demanded the ability to put up with soul-killing rejection and isolation. While our classmates were going on for MDs, JDs, PhDs, and MBAs, we would be struggling to get by on the minimum wage, performing jobs that weren't commensurate with our education to support a habit that would bring us little, if any, payment or recognition. And yet, when I told him that I had decided to become a writer, he threw back that regal white head of his and mouthed the word *yes*.

"I didn't want to tell you what to do. I didn't want you coming back here in twenty years and blaming me for ruining your life. But as long as you've decided on your own . . ." He smiled the approving smile that gives any young person the courage to spend the next twenty years withstanding the rejection and isolation of whatever calling she has pledged to follow. "Don't worry," he said. "You'll find a way to support yourself. A lot of people will want to help you along your way."

The only member of the physics faculty who asked me what I was going to do after I left New Haven was Professor Zeller. "That's terrific!" he said. "The world can always use more writers who understand physics." This was a supportive response. The world does need more writers who understand physics. If I didn't want to become a theoretical physicist badly enough to put up with the crap I would have needed to put up with, whose fault was that but mine?

And yet, if I lacked confidence, was it so wrong for me to hope that my elders might supply it? If a single professor had said, "You know, Eileen, you really are quite good at physics," I would have been quite good at physics. In fact, I would have been quite great at physics.

Of course, becoming a writer is no piece of cake. No matter how many people warn you of the hardships, you tell yourself, "That won't apply to me. I might need to wash dishes for a year, but only until I sell my first story to the *New Yorker*." Hersey announced that the English Department was sponsoring a fiction contest, and even though the faculty tended to reward their own, he urged us to apply. Buoyed by an uncharacteristic fit of confidence, I entered my story about the mime. When the day came for the award ceremony, I practically danced to Linsly-Chit, anticipating the moment my name would be called and everyone would whisper, "Who is that? She isn't an English major, is she?" "No, she majored in physics, but she decided to become a writer."

Breathless, I waited as the professor who judged the contest walked down the aisle. A jaunty, compact man with one of those impossibly cultivated mustache-and-goatee combinations, he carried a walking stick or cane, or maybe it was an umbrella with a sculpted handle, and he wore—if only in my memory—a waistcoat, a pocket watch, and elegantly tailored pants. None of this would have prejudiced me against him, but when he announced the name of the contest's winner and a jaunty, compact young man walked down the aisle wearing—if only in my memory—a waistcoat, a pocket watch, and elegantly tailored pants, I imagined grabbing his walking stick or cane, or maybe it was an umbrella, and beating him about the head with its sculpted handle.

Hearing the announcement of an award and realizing your name wasn't called is like standing at the edge of a deep gorge and knowing it's within your power not to jump. And really, I might have been strong enough to step away from that cliff, if not for the reading that followed. A famous writer had deigned to come down from Harvard and read us a novella in which he recounted the experiences of a Harvard writer who spends an hour laboring to make his frigid girlfriend achieve her first orgasm. In excruciating slow-motion detail, the story presents the real-time feelings of the man; when his girlfriend finally comes, the listener is meant to feel as satisfied as the character. What I felt was an overpowering urge to stand up and shout, "It's all about you, isn't it? A woman can't even achieve an orgasm without the help of some noble, long-suffering male writer? How come we never get to read a story about the desires of a smart Radcliffe undergraduate, written from her point of view?"

Even so, if I hadn't gone directly from the award ceremony to the dish room, I might not have plummeted into the depression that overtook me, an eddy of despair so profound I couldn't even cry out for help. In the months that followed, I would look at the dishwashing machine and feel as if my throat were stuffed with the same mishmash of garbage. What had I done? How would I support myself? If I failed as a writer, how could I go back to physics? I seemed barely able to remember how to multiply or divide, let alone take an integral. I would spend the evening working the dish machine, then trudge back to my apartment in my foul uniform, wishing I had been fortunate enough to attend class with the clever, attractively dressed young people I passed on the street.

One afternoon, I noticed a sign for a female psychotherapist. I scheduled an appointment. But when I returned, I found a woman in a stylish suit, her hair so perfectly cut and makeup so perfectly applied that I couldn't bring myself to confide I spent my days unstuffing a garbage disposal and my nights eating bags of cookies, then vomiting them back up. After fifty minutes of silence, I left and didn't return.

I was saved by winning a lottery I didn't remember entering. The semester before, a friend had suggested I apply for a Rhodes Scholarship to study in England. A Rhodes? Weren't Rhodes scholars all political science majors who had starred on their football teams and been tapped by Skull and Bones? Besides, hadn't Cecil Rhodes made his money exploiting South African miners? Well, my friend said, what about a Marshall Fellowship? I couldn't very well refuse money the British government was doling out to repay the Yanks for having rebuilt their country after World War II.

All right, I said, I would apply. But I nearly gave up when it came to composing the essay I needed to convince the selection committee I hadn't decided to become a writer merely because I had lost interest in physics. "I cannot regret following my interest in science," I wrote. "By leading a double life at Yale, where science is conducted on 'the Hill' and humanities is consigned to 'the Old Campus,' I have been able to enjoy cross-pollinating the two," as if I had just passed four carefree years buzzing from flower to flower with all the other happy, multidimensional bees at Yale. "In three years of racing from lectures on wave mechanics to ones on Martin Buber, from meetings with my advisor to discuss my research project in harmonic analysis to readings by poets such as Robert Penn Warren, I could not find the line that was reported to mark the boundary between the 'hard stuff' and 'the artsy stuff.'" Not find the line? What was that concrete barrier I kept buzzing into every time I tried to fly from Science Hill to Linsly-Chit, or from Hersey's seminar in Pierson to Peter Parker's lair deep inside the Wright Nuclear Structure Lab? But on I went, trying to persuade the Marshall committee that studying physics had been part of a cunning plan I had devised as a child, and a fellowship from the British government "would allow me to complete the bridge from physics back to writing in a way no alternative could."

When I showed up for the mock interview organized by the career center to help Rhodes and Marshall applicants prepare for the real thing, I performed so poorly—I hadn't spoken for months and couldn't gather my thoughts quickly enough to respond to my interviewers' questions about whether I admired Hemingway and why anyone would care how a wave travels in n dimensions—the director of the career center said I didn't have a prayer of winning a scholarship and shouldn't have been allowed to besmirch Yale's honor by applying.

As a result, when I received an embossed letter from the British Consulate stating that I had made it past the first stage and was requested to travel to Boston to be interviewed by a committee of seven former Marshalls, I was more apprehensive than excited. I ironed my one dress, pulled on the parka I had been wearing since freshman year (the lining was ripped, but I hadn't mended it with duct tape lest anyone mistake me for an old-moneyed Old Blue), and boarded the train to Boston. I was greeted at the British Consulate by an impeccably tailored secretary who took my parka and ushered me to the parlor, where I waited among my more poised and polished peers.

Luckily, when my turn came, the first question the panel asked was what I thought of Hemingway, followed by a request that I explain why anyone would want to know how waves travel in n dimensions. When an interviewer mentioned he could understand why a writer as eminent as John Hersey "had heaped upon me such mighty praise," I dared to hope I might progress to the final round, a hope that vanished when I stepped back into the parlor and the secretary began plucking feathers from my dress in a quantity sufficient to reclothe the flock of geese that had been sacrificed to stuff my parka.

And yet, I received a second embossed envelope, this one from the British Embassy in Washington, congratulating me on having been awarded a Marshall Fellowship to read Literature and Philosophy at the University of East Anglia in Norwich, England. (Sick of Yale's stuffiness, I had applied to one of Great Britain's newest universities, the only one that offered a degree in creative writing.) I continued to wash dishes, but the prospect of living abroad, doing nothing but reading and writing for the next two years, lifted me from my depression.

I might even have joined my classmates in their pregraduation parties, but I didn't know anyone who might tell me where the parties were. Everywhere I went I heard talk of the commencement ball I hadn't attended, the croquet match my classmates had played in nineteenth-century dress, the ivy vine they had planted while reciting odes they had composed in Latin, Mandarin, and Inuktitut. How did my fellow seniors know we were supposed to take the small white clay pipes we had been handed that morning, puff a few celebratory puffs, then trample them underfoot and lift our voices in "Bright College Years" while waving our white handkerchiefs in the air as we harmonized to the final stanza of our anthem: "So let us strive that ever we may let these words our watch-cry be, where're upon life's sea we sail: 'For God, for Country, and for Yale!'"

My parents felt even more bewildered than I did. They were proud of me, I'm sure. I had graduated summa cum laude, Phi Beta Kappa, with honors in my major. But they couldn't figure out why I had struggled to earn a degree in physics only to announce I intended to become a writer, or why they had paid tens of thousands of dollars for their daughter to attend an Ivy League university if she wasn't going to take advantage of the opportunity to marry one of her wealthy, successful classmates. I watched as my parents drove off, then looked at my diploma and realized I had been awarded a bachelor's degree in philosophy, as if whoever was filling in the blanks couldn't believe that a female student would be graduating with a bachelor of science degree in physics.

The next morning, I boarded a bus to New York so I could undergo minor surgery for a cyst on my spine. After hoisting my powder-blue plastic valise above my seat, I slid in beside the elderly black man who would be my companion. He was shabbily dressed; then again, my jeans and T-shirt hardly marked me as the product of an Ivy League university. The man asked where my family lived, and when I told him upstate New York, he said, "So what are you doing here in New Haven?" Having absorbed the unfortunate truth that if I were to admit to an ordinary mortal where I attended school, he would spontaneously combust from envy, I put my hand to my mouth and mumbled I had just gotten out of Yale.

"What's that?" the man asked. "You just got out of jail? Oh, honey, no need to be ashamed. We all done things we're none too proud of. Only thing matters now is you pick yourself up, dust yourself off, and make something of your life from here on in."

PART III

Return to New Haven

CHAPTER ELEVEN

The Two-Body Problem

Driving back to New Haven thirty-two years later, I feel as much a failure as if I had in fact graduated from the city's correctional center rather than its university. Like many women, I remember my failures more keenly than my accomplishments. Rather than take pride in having been named a Marshall scholar, I wince at the memory that I am probably the only Marshall in history who relinquished her fellowship after one year. Lonely and exhausted, I returned to the United States and moved to a bleak working-class suburb of Chicago to share a basement apartment with Laurel, who was studying for her medical degree. I was so muddled from depression that the manager of the diner where I washed dishes judged me too stupid to operate the cash register. If my former classmate at Yale, Rochelle, hadn't gotten me a job at a small but wonderfully literate newspaper in New Hampshire, which was shorthanded because of the 1980 presidential primaries, I might still be scrubbing pots and pans at the 21 Fried Shrimps restaurant in Maywood, Illinois.

The years that followed weren't easy. To fit in at the newspaper, I needed to learn to bluff my fellow reporters at poker and take ribbing from the cops. I found myself lacing my orange juice with alcohol so I could devil up the courage to pursue whatever sources I was required to interview that day. Finally, I applied to the Iowa Writers' Workshop and headed west to earn my MFA. Afterward, I moved back east to live with my future husband, who was studying for his MD at Harvard and a PhD in genetics at MIT. As Hersey had predicted, watching your classmates go on to lucrative careers in medicine, law, and business while you are temping at a sewage-supply company is demoralizing. But making it as a writer is even harder if you're a woman. The male writers at Iowa boasted of their intentions to achieve great critical and commercial success, but I never heard a female classmate voice that claim. We found it difficult to insist we hire another woman to watch our kids while we worked on stories that might not get published and wouldn't bring in more than a few dollars if they did. Obeying the cardinal rule that one should write about what one knows, I wrote my first novel about a scientist, only to be told men won't read fiction and no woman's book group would read a novel that had science in it. Struggling with my marriage—my husband rarely came home from his lab—yet deliriously in love with my newborn son, I wrote stories about being a mother and wife, only to hear male readers dismiss fiction about family life as "too domestic." I tried writing about the Borscht Belt, only to discover that a woman who writes in the vein of Philip Roth isn't "ladylike" and her bawdy humor is a turnoff.

If I didn't allow these obstacles to overwhelm me, it was because I had already tossed away one career and wasn't about to give up another. That, and I found a terrific therapist. I loved the opportunity to read and think about anything I wanted to read and think about—how consciousness and language had come to be; the ways in which space and time influence our memories; the forces that attract and repel human beings, because aren't those forces as worthy of contemplation as the forces that attract and repel atomic particles? I found myself grappling with the mysteries that had perplexed me all my life—the reality that each of us will die; that no one is more special than anyone else, although we certainly feel we are; that each of us inhabits a multiplicity of worlds, some as technologically complex as the innards of a computer, others as simple as a bubble of soap on a baby's nose.

I have never seen writing as a consolation prize. I love creating something out of nothing. I love talking inside my head. I love marking words on a page so other people can read what I write and hear my voice inside of their heads. For years, still smarting from my Orthodox upbringing, I refused to enter a synagogue. But the more I wrote, the more I thought about the theology I had learned in The Philosophy of Existence. I especially liked the legend about how God had created the universe not once but twice. The first time, he poured all his infinitely powerful light and energy into his new creation, flooding it with a oneness in which no opposites could endure. Light was a particle and a wave; space was here and there and nowhere; time was present and past and now. But the vessels of that first creation were too fragile to hold such power. Like a nuclear reactor gone critical, each vessel exploded, releasing God's divine light to propagate through whatever number of dimensions this universe really is, the sparks coming to rest in the husks of this, our fallen second world, where each of us is commanded to find the sparks of divine light that have been given us to redeem, to notice them, free them, allow them to reunite with all the other sparks and so re-create the harmony between all those false opposites, between the masculine and feminine forms of God.

Which happens to be the way a writer writes—noticing details, entering I-and-Thou relationships with every object she encounters, finding the spark in each, allowing it to fly up and connect to some other spark, uniting two dissimilar things with a metaphor that short-circuits logic and jolts the reader across the room, connecting the fragments of our chaotic, senseless life to create a meaningful thematic whole. Can writing a story bring the Messiah any sooner? I like to believe it does.

And if writing doesn't do the trick, maybe teaching will. Despite all the years I spent battling my parents' suggestion that I earn my living by teaching English, that's exactly what I have spent the past three decades doing. How could I have guessed I would derive so much pleasure from instilling confidence not only in the young women in a class, but also the gay, black, Hispanic, Native American, Asian American, and working-class young men who need it? Or I would come to love inspiring students of both genders to trust the importance of the questions that obsess them, and their ability to discover the answers to those questions? I might not be Richard Feynman, but I certainly get a kick out of taking him as my model.

And yet, driving back to New Haven to research this book, I feel like a failure. How can making up stories compare to discovering life on other planets? Maybe everyone who graduates from Yale feels that whatever he or she has achieved can't compare to what his or her classmates have accomplished. Or only the women feel that way. Maybe that's why the ophthalmologist I met at my class reunion, the woman who stunned me by revealing she had graduated with a BA in physics the same year Erika and I earned our degrees, declined to respond to my repeated attempts to contact her. Or why the first thing my friend Leslie said after I had tracked her down to interview her for this book was, "Oh, Eileen, you don't want to talk to me! I'm such a disappointment!"

In New Haven, I check into a motel beside the highway, then obsess about what to wear for my interview with Michael Zeller. I put on one outfit, decide I look too sexy and unprofessional, change to a suit I reject as frumpy, then pull on jeans and a baggy shirt, as if I am still eighteen and trying to prove my seriousness as a physicist. I drive to campus and wander the dreary chemistry building in which I once tried to replicate the Millikan oil-drop experiment, then descend into the tunnels through which Greg and I ferried our mystery solutions to the pre-med students, ending up in the subterranean reading room of Kline, where five trollish young men stare at their books without seeming to notice the buzzing fluorescent tubes overhead. I am dizzied by the thought of how much time I wasted down here working on problem sets that brought me As in courses I never used because I didn't go on in physics. Then again, I remember the glimpses of heaven I was granted in this same basement and wonder what it would have been like to spend the previous thirty years experiencing such exhilaration on a daily basis.

For years I have wanted to thank Michael Zeller for not allowing me to drop his class, and I can't wait to show him the brittle fanfold pages of the computer program whose Xs and Os mark the simulated paths of the particles he once smashed against each other at Brookhaven. "Of course I'd like to talk to you," he wrote me. "A lot has changed with respect to women in physics. Our Chair is now a woman." He had just retired, but would be happy to arrange for me to talk to current students and female faculty. "I look forward to seeing you after 35! years."

I know I shouldn't be picturing him as the broad-shouldered young man whose charismatic lecturing style drew me to Yale. Then again, he must have retained some of his old appeal: I turned up an article titled "Zeller's Laws of Attraction" in which a female reporter from the *Yale Daily News* queried my former professor not about the changes in physics in the past half a century, but the changes in the interactions between the sexes. The interview, published in 2007, makes me realize that even though my professors seemed young and hip to me, they had formed their ideas about women in the fifties. "I was there at the dinner table," Zeller told the reporter, "but mostly my children were raised by my wife." Not that he wasn't thrilled that women now have opportunities they didn't have in his day; he was just glad he and his wife didn't need to figure out how to make a two-career marriage work. Citing a young woman to whom the Physics Department recently offered a job that put her husband in the position of choosing between his career and his wife, Zeller joked, "They've got what we call a 'two-body problem,'" then asked the reporter to excuse his physics humor. But how could I not love a guy who described his wife by saying that after forty-six years of marriage, "she's become a leg. I know everything about my leg and I enjoy having it and I wouldn't want to live without it."

Sadly, this article turns out to provide the only insights into the mind of Michael Zeller I am going to receive. After waiting hours, I get a message in which he apologizes for missing our interview on account of a family emergency. Later, I learn he recently suffered a stroke. A stroke? How could such an infirmity have befallen the bright-eyed young man who raced across the stage explaining centrifugal force with such enthusiasm? Suddenly, the questions I have come to ask seem pitifully insignificant compared to whether my professor will be allowed to enjoy the company of his wife for decades to come, as she will continue to enjoy his.

. . .

To interview Peter Nemethy, I take the train to Manhattan, where he teaches at NYU. I am too shy to ask, but I can't help wondering if he failed to get tenure at Yale because he persisted in attending concerts and spending enough time with his wife so he remembered what she looked like. In his photo on the NYU website, he appears remarkably youthful, but in person he looks his age; his eyes, which once projected joy at the physics he was teaching, now project a profound melancholy. I dismiss this as a product of my imagination—until Peter tells me that his wife died not long before. She was a PhD chemist, he says, which is why it always felt totally natural to him to have women in his physics classes. Of all my professors, only Peter doesn't remember me, but that might be because it never struck him as odd that I was there.

"I never felt 'gee whiz, why is a girl here,'" he says. What bothered him was how long it took American universities to start hiring female professors. Even today, the ratio of women to men is "too small by a lot," especially in particle physics—an imbalance he is quick to point out is not true internationally. CERN, where Peter carries out his research, involves the collaboration of thousands of physicists representing hundreds of universities, and yet female scientists are not underrepresented, even at the top. His boss at CERN is an Italian named Fabiola Gianotti who "holds the whole group together, and not a single person at CERN says, 'Gee whiz, why is this job being done by a woman.'" Countries like Italy field research groups that are fairly well balanced between women and men, while the groups from the United States and Germany are comprised mostly of men, a disparity Peter is at a loss to explain.

Well, I say, it doesn't help that American men don't want to date girls who are good in science and math.

He grins a grin that reminds me of the old Peter, or maybe what I mean is the young Peter. "I *have* heard that in America, in high school, it's not considered in dating to be an asset. And that is something I have never heard in other cultures, that you wouldn't want a woman to be strong in mathematics, to be strong in using her mind." When I venture that girls are more inclined than boys to see a low score on an exam as an indication they ought to switch majors, he wags his head noncommittally. He encounters many students of both genders who get a bad grade and give up. He's just not sure if that's because they are too easily discouraged, or because they shouldn't be in physics. "What I've seen in a few cases are people who struggled very hard in the first year but persevered, like you did. They decide they'll push on in spite of that and end up doing very well." He contemplates his knees. "Should I provide those students with encouragement? I can help students with the logic of 'how does one approach things.' But I don't know if I can provide mentoring at the emotional level."

He tells me that his first PhD student at Yale was a woman, and she was wonderful in both the classroom and the laboratory, which made him find it "totally natural" that she went on in physics. He had a female graduate student at NYU whom he discouraged from going on, but he's not sure he can quantify why. A student's performance in the classroom isn't very predictive of his or her success as a practicing physicist. At the earliest opportunity, anyone interested in going on in science ought to be encouraged to do research, Peter says, not only because lab experience will help you get into graduate school, but because it will help you figure out if you are going to be good beyond the classroom. What Peter looks for in a research assistant is a combination of being able to work well independently and yet not being a loner. "If somebody is very insecure on a project and needs to be told what to do at every step of the way, that to me is a bad sign," a pronouncement that makes me cringe at all the times I got stuck on my project and needed to visit Roger Howe for help.

I ask what Peter does when he encounters a student who seems to have the boldness, independence, and creativity to go on in physics, reminding myself this is one professor I can't blame for not having encouraged me to go on. After all, I asked him to write me a letter of recommendation for graduate school after only two years at Yale. The advice he gave me—that I should follow his example and lead a balanced life—was advice I needed to hear. He wasn't my father. He had no obligation to follow up and inquire why I never did ask him to write that letter.

Besides, I find out now he never encourages any of his students to go on in particle physics. Being successful requires incredibly hard effort, he says. You have to travel, and when you get time on an accelerator, you need to let your experiment take over your life. "Anybody who is only sort of interested but not driven to it would have a hard time." And he maybe gets the sense women are not as driven as men. "Over the years I have found very many wonderful women students. I have written wonderful letters for them. I have had discussions with them about where they ought to go. But the sense that I should give them a validation to go on, that is not something I ever felt the need to do." He frowns. "It's not at all true the male students did better. But even with what you say, I would still be hesitant pushing *anyone* into physics."

I ask if he still thinks it's possible for a physicist to be successful without giving up everything he loves to pursue his passion. No, no, he says, jerking his head. He has seen colleagues who gave up everything for physics, and others who didn't, and both kinds turned out to be successful. As for him, he decided early on that giving up everything he enjoyed would be too big a price to pay, and if he didn't end up being quite as successful as otherwise might be the case, so be it.

Still, the traveling took its toll. He and his wife decided not to have children because his wife was afraid he would be an absentee father. I'm too tactful to ask if he doesn't regret that now—even with his wife gone, he might have their children to keep him company. Then again, he has his students at NYU. He has those thousands and thousands of colleagues at CERN.

Before I go, Peter grins and says he has a question for me. If it's true that his advice led me to become a writer, and if it's true that he once predicted I would dedicate a book to him . . .

Blushing, I reach inside my pack and bring out a copy of my latest story collection, which I already have inscribed to him. I want to give him a hug. But he still strikes me as too formal, and I content myself with noticing that the sadness in his eyes has given way to pleasure, and he seems sorry to see me go.

· · ·

To keep my appointment with Peter Parker, I approach the tomb-like Wright Nuclear Structure Lab, set like a pyramid into the base of Science Hill. I expect to be confronted by a sphinx demanding I answer a riddle about nuclear reaction rates, or at least show an ID. But no one questions my presence, and I wander the corridors looking for an office with a giant spiderweb across the door.

"Eileen? Is that you?" The man who says this resembles the Peter Parker I used to know, except his black hair and black beard seem to have been bleached white by a blast from the Van de Graaff accelerator. As this new/old Professor Parker leads me to his bunker (the walls, I am disappointed to note, are nearly devoid of Spider-Man paraphernalia), I remember how much I like the guy. He is as unguarded as the entrance to the reactor. He doesn't give off the slightest whiff of ego.

My fondness might also have something to do with his remembering who I am. Although what he remembers isn't my status as the first female undergraduate he ever taught, but an incident I recounted in the *Scientific*. "You're the one who told the story about the toilet!" Peter says. I laugh and admit I am. In one column, I described the time I summoned my father to fix our toilet, and he said, *Eileen, if the mind of man invented it, your mind can figure out how to fix it.* "I've been telling that story ever since," Peter says, and I don't get the idea he tells it more frequently to women than to men.

We settle in our chairs, and I feel as if I am eighteen, here to ask my advisor which courses to take. Except this time, I tell the truth about how abysmally far behind I was when I got to Yale, how close I came to dropping the major. Peter nods—it's a story he has heard dozens of times, not only from women. Every summer, as director of undergraduate studies (DUS), he goes through a list of all 1,300 members of the incoming class and, using a combination of test scores and transcripts, sends each one an e-mail advising him or her which courses in science and math to take. Surprisingly, half the students who declare a major in physics come from the nonintensive track. And by junior year, when Peter teaches the intermediate course in quantum, he can't tell the difference between the kids who arrived with great preparation and those who arrived with a lousy background. Like me, lots of kids hit their first problem-set or exam and find themselves in trouble. "They realize you can't just plug numbers in equations the way you did in high school. You have to understand what you're doing. We try to make sure that the people teaching our intro courses are tenured profs, the best teachers we have, teachers who are prepared to put in the time. We've added tutoring sessions since you were here. We encourage the students to form study groups. The trouble is, the ones who seek out help aren't necessarily the ones who need it most. It's hard to convince Yalies they need help."

When I describe the horrors I experienced in my lab course, he laughs and says that must have been when the equipment was still on the top floor of Sloane, with bats to keep the students company. Even he couldn't get the Millikan oil-drop experiment to work. The labs are better now. The equipment is better. But the biggest change is that the department places far more emphasis on real research. All undergraduate majors are required to spend at least one semester carrying out a project in a professor's lab, and they are encouraged to do research over as many summers as possible. What Peter looks for in an undergraduate is the ability to overcome obstacles. A student shouldn't need to come in on Monday and ask, "What should I do?" and be told, "Do this," then come back on Wednesday and ask, "What should I do?" and need to be told, "Do that." The student should be able to talk to people in the lab and get opinions on how to approach the problem, then take a fresh look and show the insight to think in new ways.

When I ask what makes Peter consider an undergraduate good enough to succeed as a theoretician, he shakes his head, as if I had asked him to define the qualities that might allow a Yale drama major to grow up to be Paul Newman (class of 1954) or Meryl Streep (1976). If a professor takes you on as an advisee in theoretical physics, he will expect you to be good and bright. "And you need the math. You can be a clever experimentalist without knowing group theory and differential geometry, but not so if you're going on in theory. For the experimentalist, math is a tool; for theoreticians, math is almost another love." It's also much harder to find a job as a theoretician. These days, most PhDs in theoretical physics end up working on Wall Street.

I nod—a lot of people have told me that theoreticians find their skills valued more highly in the financial sector than in academia. I ask if he still holds a yearly meeting to help graduating seniors evaluate their career options. Oh, no, he says, he leaves that to the Society of Physics Students. But as DUS, he does meet individually with each senior physics major. He asks, "Where do you see yourself a year from now?" And if someone tells him, "I'm applying to medical school," or, "I'm going on in oceanography," he doesn't say, "Gee, you really ought to consider physics grad school." If someone were wavering and asked him for advice, if someone were to say, "I don't know if I'm good enough," then sure, he would encourage that person. Otherwise, he doesn't actively encourage—or discourage—anyone.

I ask how many female undergraduates he has mentored, and Peter says that in his four decades at Yale, he's had only a few undergraduate research assistants. Of the thirty or so doctoral candidates he has supervised, only two or three have been women, and those in the past few years. Still, he says, those female students were great. Of the six or seven top physics students in the present class, the three or four with the highest grades are women. I ask why, even with the increase in female undergraduate majors, so few women remain in physics long enough to achieve tenure.

Peter gestures around his windowless office, piled high with equipment in stages of disrepair. "This is what defines me," he says. "I love what I do. I'm prepared to give up on the outside world. But we still live in a society where childrearing is mostly a female responsibility. The culture is gradually changing. But that's still the way it is." Then again, all those slow changes have added up to a revolution. As an undergraduate, he attended an all-male school. He can't recall if Caltech, where he earned his graduate degree, was limited to men, but there were no women in his class. "It didn't even seem strange. And I look back with horror now that it *didn't* seem strange. It used to be a scientist considered a woman someone with whom he could spend a Friday night date, relaxing, whereas now he sees a woman as someone he can have a professional, intellectual relationship with. Someone to study with. The person at the opposite desk who solves problems better than you can. It's a huge difference, and God bless it. I find it hard to imagine how things used to be."

Peter also attributes the change in the way male physicists see their female colleagues to the role men now play in their children's lives. He once went to see the guidance counselor at his daughter's school, and he noticed the posters showed white-coated men as doctors while the women were holding clipboards. "'What is wrong with this place?' I thought. Anyone can tell you, the one way you cure a male chauvinist is to give him a daughter."

I save Roger Howe for last. "How nice to hear from you!" he wrote in response to my e-mail. "I do remember you, and I tried to find out where you had gone after graduation, because I wanted to let you know about a sequel to your project. But the Physics Department couldn't give me any contact information." He seemed enthusiastic about my book, adding that the Physics Department now has a female chair and she is "a fierce advocate" of women.

We made an appointment to meet in his office—I joked that I wanted to compare its level of messiness to the way it looked thirty-five years ago, and he warned me the new office, which used to belong to Professor Kakutani, is more of a mess than the old, but I never get the chance to check because when I arrive—admittedly, a few minutes early—the door is closed, and I hear a familiar voice apologize, "I'm not really dressed." The door opens a crack; I glimpse my former advisor in a T-shirt and sweatpants, and he asks me to wait while he puts on something more appropriate. Oh well, I think, he must have come straight from the squash court, and I use the time to climb to the top floor to use the ladies room; someone has tacked a poster of "Famous Women in Math" beside the restroom, but the larger poster of famous male mathematicians is still given pride of place on the main floor.

Roger's door remains shut, so I listen to a calculus review session led by a graduate student with long, tangled hair who stands at the board writing in barely legible numbers and offering an explanation so unilluminating I could do a better job despite not having taken a math course since 1977. Roger emerges in a button-down blue shirt, khakis, and sunglasses. Despite his assertion that he is too old to play squash, he appears remarkably youthful, even when you consider that when I studied with him, he was the youngest full professor at Yale. He might be a little more gaunt and ethereal, but by any standards he is still a strikingly appealing man.

He suggests we grab a sandwich around the corner, and as we sit waiting for our paninis, I ask about his childhood, expecting a story about how, like Jesus among the elders, he startled his teachers with his precocity. In fact, Roger's childhood seems oddly like mine. Or rather, my childhood if I'd had a father who was a chemist for the Manhattan Project and encouraged my scientific curiosity at every step. Roger grew up in Ithaca, which resembles my hometown in upstate New York, except instead of decaying Borscht Belt resorts and a community college designed to train students for the hotel industry, Ithaca can boast two major universities. In fifth grade, Roger's father bought him a set of Dover math books, and he demonstrated enough aptitude that his father told him that he was smart, his teacher predicted he would grow up to become a mathematician, and his principal saw fit to skip him from fifth grade to sixth.

In tenth grade, Roger found a calculus book and, like me, thought the subject was "really cool." At the start of his senior year, he solved a problem no one else in his class could solve, and the teacher sent him to Cornell to find a course more in keeping with his abilities. But even at Harvard, he wasn't immediately recognized as a genius. He was told not to take the most rigorous math class and advised he wasn't ready to take advanced chemistry because he "didn't have the math." But Roger was confident enough to ignore what people told him. He sat in on the more advanced classes, figuring he could do whatever math the course required. There might have been a few zigzags, but Roger's talents were quickly recognized and he was sucked into the world of upper-level seminars, research projects, and competitions that define the career of a top mathematician. After earning a doctoral degree at Berkeley and teaching a few years at another college, he arrived at Yale in 1974, the same year I did.

When I tell him one reason I didn't go to graduate school was that I compared myself to him and judged my talents wanting, he shakes his head and says, "But you were an undergraduate!" I want to ask if he thought I was any good at math, but I am afraid of what

I might hear, so I come at the question more obliquely. "I hated the way I had to keep asking for help. I hated not making any progress for so many months."

He looks puzzled. "But you solved it."

"Yeah," I say. "At the end I really understood what I was doing. But it took me such a long time."

"But that's just how it is," he says. "You don't see it until you do, and then you wonder why you didn't see it all along."

"Really?" I say. "I got discouraged because I needed so much time to figure out what you already knew. And I couldn't finish Professor Kakutani's course in real analysis."

Roger shrugs. There are a lot of different math personalities. Different mathematicians are good at different fields.

I use this as an opening to ask if he ever noticed any differences between the ways male and female students approach problems, whether they have different "math personalities." No, he says. Then again, he can't get inside his students' heads. He did have two female students go on in math, and both have done fairly well.

That's great, I say, then ask why even today there are no female professors on Yale's math faculty. No *tenured* women, Roger corrects me. Just recently, the department voted to hire a woman for a tenure-track job.[1] Well, that's still not very many. Why has it taken so long to hire even one female mathematician?

Roger stares into the distance. "I guess I just haven't seen that many women whose work I'm excited about." I watch him mull over his answer, the way I used to watch him visualize n-dimensional toruses. "Maybe women are victims of misperception," he says. Not long ago, one of his colleagues admitted that back when all of them were starting out, there were two people in his field, a woman and a man, and this colleague assumed the man must be the better mathematician, but the woman has gone on to do better work.

Hearing all this upsets me so much I come straight out and ask Roger how my project compared to all the other undergraduate research projects he must have supervised since the seventies.

---

1. Later, the math faculty also hired a tenured female professor.

His eyebrows lift, as if to express the mathematical symbol for puzzlement. Actually, he hasn't supervised more than two or three undergraduates. "It's very unusual for any undergraduate to do an independent project in mathematics." He pauses. "By that measure, I would have to say that what you did was exceptional."

"Exceptional?" I don't know whether I want to lean over the table and kiss him, or wrap my hands around his throat. I ask why he never told me this before.

The question takes him aback. He guesses he didn't used to think in those terms, which either means he was more concerned about his own career than his students', or he didn't think about encouraging women to go on in math. I ask if he ever tells any of his undergraduates they ought to go on for their PhDs; after all, he is now the director of undergraduate studies. But Roger says he has never encouraged anyone to go on in math, and probably never will. "It's a very hard life. You need to enjoy it. There's a lot of pressure being a mathematician. The life, the culture, it's very hard. Teaching at a place like Yale, a mathematician has so many pressures and responsibilities, even apart from the pressures of carrying out one's research."

We sit in silence until I ask why he contacted the Physics Department to get in touch with me. Visibly relieved, he pulls out a paperback titled *Non-Abelian Harmonic Analysis: Applications of SL (2, R)*, which he coauthored with a mathematician named Eng Chye Tan. (The book's ranking on Amazon, which I later discover to be 7,715,384, makes me feel better about my own books' numbers.) Roger shows me the five pages taken up by the group theoretical proof of Huygens' principle and apologizes for not having acknowledged my contribution to the work. I tell him that I don't mind (I am tempted to say all I did was take dictation). But deep inside, I wish I might have left even the tiniest footnote in the world of non-abelian harmonic analysis.

A colleague stops by, and I sit wondering what would have happened if Roger had told me all those years ago that my ability to solve his problem was exceptional. A math professor at the University of Michigan once assured me a recommendation from Roger Howe, combined with a contribution to a published paper, would have placed me at the top of any graduate school's pool of applicants.

Would I have gone on to produce work my former advisor found exciting? Would I have been willing to disappear down the same rabbit hole of mathematics as Roger and this shaggy, pot-bellied colleague? The Eileen I am now recoils in horror. But I never would have become the Eileen I am now if I had gone to graduate school in theoretical physics. And the Eileen I used to be . . . that Eileen would have been incredibly proud to have her name in the book she is holding in her hand.

Roger puts on his sunglasses and we stroll back to Leet Oliver, where we exchange heartfelt expressions of how nice it was to see each other again. And I mean what I say. As much as I wish he had encouraged me to go to graduate school, if not for Roger Howe, I wouldn't have been so captivated by mathematics in the first place. I wouldn't have had the abysmally frustrating and exhilarating experience of solving a problem no one before me had solved. In a column I wrote for the *Scientific*, I described my senior research project as "an act of communication." I came to understand a thimbleful of another person's thoughts. I accomplished the sort of mind meld I envied Mr. Spock for being able to carry out on *Star Trek*.

"So." Roger lifts his sunglasses. "I guess what you're saying is that women are more . . . more . . . *discourageable* than men?"

Well, no. That's not how I would have put it. But I want to reward the effort. "Sort of," I say, then kick myself for leaving him with the impression that women are even more deeply flawed than he might have realized.

As everyone keeps reminding me, 30–40 percent of the undergraduates majoring in physics and physics-related fields at Yale are now female. Surely these women must be confident and tough and far less in need of encouragement than I was.

To check out this hypothesis, I sit in on the intensive introductory physics class I would have loved to sign up for as a freshman. And sure enough, the diminutive young woman in the pink sweater and tiny brown-rimmed glasses who is sitting in the empty lecture-hall typing diligently on her laptop when I arrive strikes me as the sort of student who didn't exist when I went to school here. She studied calculus in eighth grade, then physics in ninth, followed by AP chem and bio, leaving her with nothing much to study her senior year, "which was okay," she says, "because I needed a break to apply to colleges." After a gap year in Israel, where she attended an all-girls school ("It was good for me. If I got stuck on something, I couldn't ask a guy for help"), she arrived at Yale and signed up for the intensive track in physics, along with Real Analysis. "I'm doing okay," she says modestly about the only course I needed to drop.

By now, the auditorium is nearly full, and of the fifty students, twenty are women. As my informant in the pink sweater points out, the male-to-female ratio in physics isn't bad. The problem is in her math class, where she's the only woman. "In math class," she says, "I can't relax and sprawl out. I need to maintain some dignity."

The lecture is about to start, and even though I used to sit near the front of any class so I could concentrate and take notes, I climb to a seat in back. Schools such as Harvard, MIT, the University of Colorado, and the University of California, Davis, have taken to more interactive, student-centered methods of teaching physics, but nothing much has changed at Yale since I walked into Michael Zeller's lecture in the seventies. Still, today's lecture on collisions in two dimensions is clear and engaging. The professor's attempts at humor are clever, and none of his illustrations require a familiarity with football or warfare; to introduce the formulas for adding up the individual bits of angular momentum in a spinning tire versus a spinning disc, he employs the inspired metaphor of bees flying about a person's head, individually or in a swarm. I lean back and relax, thrilled that a class I never could have passed as a freshman looks so easy now. If only I had attended a better high school, I wouldn't have had any trouble acing the intensive sequence. Maybe, like the young man beside me, I would be eating soup and muttering, "Jesus, I *know* all this already."

I look around and realize far more women than men are taking notes, whether because they are not as familiar with the material or because they are more willing to admit they might need the professor's insights to figure out that week's assignment. And when the professor asks if anyone has any questions, more women raise their hands; again, this might be because the women are less prepared, or because the men are trying to project omnipotence. I am tempted to ask a question about angular momentum I have always been embarrassed to ask because the answer seemed so obvious, but I don't want to call attention to my presence as the only fifty-four-year-old in the room.

As if she can read my thoughts, the girl in the pink sweater raises her hand and asks exactly the question I am pondering. "That's a good question!" the professor says, and then he explains the answer, which turns out not to be so trivial.

Walking back up Science Hill to meet with Meg Urry, the chair of the department in which I used to be a student, I am prepared to believe that in my decades-long absence, Yale has become a utopia for young women majoring in physics.[2] In an earlier exchange, Urry predicted the female students would recognize the issues she and I faced, "but they have a network and a support system (30–40% women among our majors) that make (I think) all the difference." Since becoming chair, she has instituted nightly study halls, which provide a structure for students of both genders to work with physics TAs to understand and complete their homework. Under the direction of a female professor named Bonnie Fleming, the students formed a group that sponsors a semiregular Conference for Undergraduate Women in Physics at Yale (like me, Urry finds the acronym CUWPY to be ironic because the women don't seem to know that a Kewpie is a naked, smiling doll awarded to men at carnivals). Both of us assume we will spend our time trading war stories and marveling at how much has changed since the bad old days. "You and I are exactly the same generation," Urry tells me in her e-mail, speculating that the only reason she went on to graduate school and I didn't is that she attended a less intimidating college—Tufts—and had a more supportive family.

From her CV, I know that after leaving Tufts in 1977, Urry earned her PhD from Johns Hopkins. She completed a postdoc at MIT's center for space research and served on the faculty of the Hubble Space

---

2. I also am amazed at the similarity between this physicist's name and "Meg Murry," the name of the protagonist of one of my favorite childhood books, Madeleine L'Engle's *A Wrinkle in Time*.

Telescope before Yale hired her in 2001 as a full professor. As an expert on supermassive black holes, she has published an astronomical number of articles in prestigious journals. What is less obvious from the CV is that in recent years Urry has become devoted to using logic, hard data, and anecdotes from her career to alter her colleagues' perceptions as to the causes and effects of the paucity of women in the hard sciences.

On February 6, 2005, in response to the Summers controversy, Urry published an essay in the *Washington Post* describing her own gradual realization that women were leaving the profession not because they weren't gifted but because of the "slow drumbeat of being underappreciated, feeling uncomfortable and encountering roadblocks along the path to success." Having come of age in the 1960s, when discrimination against women seemed a thing of the past, and having been assured she would have no trouble getting ahead because universities and research labs would be eager to fill their affirmative action quotas, Urry at first interpreted her own repeated failures to be hired or promoted as proof she wasn't good enough. Like me, she assumed it must be her fault that her suggestions at meetings went unheard, even though the men who offered similar suggestions received vociferous praise. Only as she grew older did she see that the real problem lay in her willingness to believe the negative messages she was receiving. She has no issue with women leaving science to raise a family. What troubles her is that women may use "family" as an excuse to abandon their careers when in fact they have been driven out. Male physicists assume anyone who leaves science does so because she is weak; Urry believes the women leaving the field are some of the best scientists we have trained.

As for her, she loves her job at Yale, and if some people want to think she got her professorship because she's a woman, "I'm finally able to say, confidently, that I'm really great at this job. I'm lucky to be here at Yale, yes, but even more, they are really lucky to have me. The doubt is finally going away."

Anyone who meets Urry now would have a hard time seeing her as lacking in confidence. She has the quizzical smile and radiant eyes of Jodie Foster (magna cum laude, Yale class of 1985), but a Jodie Foster who majored in physics and stuck around New Haven long enough to become chair of the department, give birth to two daughters, and switch from shoulder-baring gowns to corduroy skirts and unstructured jackets. She carries a giant purse in which she rummages for her date book or cell phone or the journal article she wants to show you, and not one but five people describe her to me as the busiest woman on campus. Although she has promised me an hour, I waste thirty minutes waiting at the wrong office (Urry has two), and by the time I find the right one, the men in the research group she supervises are clamoring outside her door.

In the little time that remains, she suggests that with so many women now studying physics at Yale, and so many of them at the top of their class, the faculty can't help but be educated that there's no difference between the abilities of their male and female students. She tells me that she was slightly disturbed to discover that 100 percent of the entering graduate class that year—about a dozen students—are men, but the statistic strikes her as a fluke.[3] She is not sure she will be able to attend the tea being hosted for me that afternoon by the master of Silliman College, an event hastily put together in the hope of attracting a few female science majors willing to let me interview them, but if she can squeeze in the event, she will be there.

Even though I lived in Silliman for three and a half years, and even though the master of Silliman held teas for us every term, I need three tries to locate the entrance to the master's house, having been too intimidated by someone called "the master" hosting something called a "tea" to have attended such events (except once, when I caught wind that Paul Newman would be the guest). I don't recall the Old Blue who was master in my day—the photo in my yearbook shows a bespectacled legal scholar in a plaid sports coat and clashing diamond-patterned tie—but if the current master, Judith Krauss, had held the position, I might have stopped by more often. "Master K" is

---

3. In the 2013 incoming class, eight of the twenty-six incoming graduate students were female.

the former dean of nursing. Older than I am, with bristly hair and a puckish face, she reminds me of Mary Martin playing Peter Pan. Her husband is a nurse midwife who holds a divinity degree from Yale; together, they are the biological parents of two grown daughters and the stand-in parents for 450 Sillimanders of both genders.

Master K warns me not to expect more than half a dozen students for the tea; the only publicity she had time to rustle up was a notice in the *Yale Daily News* that I would be on campus to discuss "gender inequalities in science." Oh, don't worry, I say, thinking that if five or six young women show up, I'll be delighted.

And so, when eighty young women (and three curious men) crowd into the dining room to pick up their cookies and tea, then file politely into the parlor to find places on the chairs and floor, Master K and I are stunned. By the time Meg Urry hurries in, she is lucky to get a seat.

I listen to Master K describe my achievements, and I feel as if everything I have accomplished has been part of a far-reaching plan in which I majored in physics to gather data on why women don't go on in science, then became a writer so I could write up what I learned. After relating a thumbnail version of my story, I ask the students to compare my experiences to theirs. I suspect they will stare at me blankly, or ask a few dutiful questions about what life at Yale was like in the bad old days. Instead, one young woman volunteers that she was disconcerted to find herself one of only three girls in her AP Physics course in high school, and even more upset when the other two women dropped. Another student says she was the only girl in her AP Physics class from the start. Her classmates teased her mercilessly. "You're a girl, we don't need to listen to you. Girls can't do physics." She kept expecting the teacher to put an end to the teasing, but he didn't. Other women say their teachers were the ones who teased the most! In one physics class, the teacher announced the boys would be graded on the "boy curve" while the one girl would be graded on the "girl curve"; asked why, the teacher explained he couldn't reasonably expect a girl to compete in physics on equal terms with boys.

And no, it still isn't cool for a girl to be too interested in or adept at science. At fancy private schools, it isn't even cool for a boy to be interested in science. (Later, browsing the Yale bookstore, I read in the latest edition of *The Preppy Handbook* that "preps don't speak physics.") When I confide that I used to be self-conscious crossing my legs in class, the girls murmur in recognition. Even at Yale, they are reluctant to ask questions because "boys give off the aura of knowing everything."

The only members of the audience who don't know what the rest of us are talking about are the women who attended all-girl high schools. One self-identified jock says she is sure all this stuff was going on around her but she was so into sports she was oblivious. A student from Spain is dumbfounded that American men don't want their girlfriends to be smart. If a Spanish girl is smarter than her boyfriend, the boy brags about how smart his girlfriend is. The American-born students shake their heads. Never mind Spain, their classmate must have grown up on Mars.

One woman—I take her to be Indian or Pakistani—says she got a great education growing up abroad. She arrived on campus having taken lots of advanced classes and didn't hesitate to sign up for the most rigorous math course. A bit shaken to find herself the only girl, unable to follow much of the initial lecture, she asked the professor: Should I be here? *If you're not confident that you should be here—* she imitates his scorn—*you shouldn't take the class!* She might have followed his advice, except she talked to the boys and they admitted they couldn't follow the lecture either.

A student peeks in from the foyer to ask whether the women in my day were reluctant to call themselves feminists. "Whenever anyone raises her hand, she always prefaces it by saying, 'I'm not a feminist, but . . .' It's as if everyone is afraid that if she's perceived to be a feminist, the boys won't ask her out." Oh no, I think, I can't be hearing this at Yale. Not in 2010. But the students nod in agreement.

Master K thanks me for inspiring such a "surprising" discussion, then presents me with a scarf printed with the Silliman flag and an engraving of the college, and though the scarf is fairly hideous, and the engraving shows two men in trench coats and fedoras crossing the courtyard, I find myself tearing up, this being the first time I have ever felt I am a true Sillimander. Or went to Yale. Or accomplished anything since I left.

A dozen girls stay to talk, all of them attractive and fashionably dressed. "The boys in my group don't take anything I say seriously!" an astrophysics major complains. "I hate to be aggressive. Is that what it takes? I wasn't brought up that way. I'm worried. Will I have to be this aggressive in graduate school? For the rest of my life?"

One striking young woman says she hates when she and her sister go out to a club and her sister introduces her as an astrophysics major. "I kick her under the table. I hate when people in a bar or at a party find out I'm majoring in physics. The minute they find out, I can see the guys turn away."

"Even at Yale, the boys won't date a physics major!" her friend wails. "I don't want to go through four years here without a date!"

I assure them finding men who enjoy the company of smart women becomes easier the older you get. My ex-husband was overjoyed to find a wife who understood his fevered descriptions of what he had accomplished in his genetics lab. And the man I dated for ten years after my divorce professed it was a turn-on to get into bed with a full professor. (What I don't say is that I divorced my husband largely because he never came home from his lab and left nearly all the housework and childrearing to me, or that my boyfriend was Polish, and Polish men, like their Italian and Spanish counterparts, find brainy women to be sexy in ways most American men don't.)

After the students leave, I ask Meg and Master K if they are as flabbergasted as I am. "More!" Meg says. After all, she is the chair of the department in which most of these girls are studying! All three of us shake our heads at how little has changed.

"Come to see me again tomorrow," Meg orders. "I'll clear away all my meetings. I think you and I need to sit down and have a very long talk."

CHAPTER TWELVE

Statics and Dynamics

After that, Meg spends the rest of the week making sure I get to talk to this or that faculty member, graduate student, or postdoctoral fellow. I go back to Ann Arbor and compare the experiences of young women at the public university where I teach to those of the students at my more privileged alma mater. I even return to my hometown in upstate New York to see what has or hasn't changed since I attended school there. Wherever I go, I am startled to discover that even as a few more girls have come to occupy the seats in math and science classes, and even as a few more women have come to stand before those classes to teach, relatively little has changed beneath the surface, especially as young female students progress from high school to college, from undergraduate to graduate studies, or from graduate school to a job.

As I push open the heavy glass doors of the building in which I attended kindergarten through sixth grade, I can hear the winds of time rush past. I haven't been inside since 1968, but nothing much has changed except the guard booth. My town is poorer than it used to be—70 percent of the students now receive free or reduced lunches. Maybe the guard booth is a sign that third graders pack automatic weapons. But I am suspicious of any claim that human nature has changed so radically in the past few decades. More likely, we have simply become willing to recognize the poverty, violence, and sexual predation we once ignored.

The principal, Jeri Finnegan, is a cheerful woman my age with a round face and hair the dull red of a penny. When I remind her that I am interested in finding out what her school might offer a child as bored as I was, I can guess from her expression how irrelevant the question seems. Mostly, she is concerned with the students at the bottom. Every once in a while, she skips a child. But those kids tend to be older than their peers, more mature and well socialized, so they will fit in with their classmates a grade ahead and not stick out like sore thumbs. (I wince, thinking of Mr. Spiro's assessment that I wasn't mature or well socialized enough to be skipped, and wondering if the brightest kids won't always stick out.)

As to science, Finnegan shakes her head. Only half an hour every other day is devoted to the subject, and even then, the curriculum is less than innovative—the same lima beans sprouting in mayonnaise jars, the same celery stalks propped in ink. Science is the least comfortable subject for most teachers, Finnegan says. With rare exceptions, the teachers fear they won't be able to answer the students' questions and are more at ease teaching from books than doing hands-on lessons. In elementary school, girls are just as interested in science and science fiction as the boys—the girls especially like Harry Potter. But by junior high, where science consists of performing experiments rather than reading books, the girls lose interest. They get scared away by the math. It's related to their home life. The guys are taking things apart, helping their father fix the plumbing and work on cars. "I didn't get a microscope for Christmas. My brothers did. I didn't get an Erector Set—my brothers did." So why should it come as a surprise that her brothers grew up to be more into technology and math than she is? It's a self-perpetuating cycle. When her daughter was in fifth grade and needed help in math, Finnegan was at a loss.

I ask if elementary teachers might at least try to get their students to appreciate how incredibly cool and complicated the human body is, how amazing it is that the planets and stars exist, that animals and plants can do all the neat things that animals and plants can do. "I wish," Finnegan says wistfully. "But with everything else we need to do . . ." She imitates a stressed-out teacher, out of her depth with science. The best the school can do is sponsor assemblies, she says. They once brought in Bill Nye the Science Guy. The librarian reads the students *The Magic School Bus*. The math units stress not only memorizing number facts but manipulating rods and blocks so students can get an intuitive feel for what the numerical operations represent.

When I ask if the district has discontinued its field trips to the Museum of Natural History, Finnegan surprises me by saying the PTA still raises money to take the students on at least one major trip a year. In the higher grades, teachers take their students to the zoo or, yes, to the Museum of Natural History and the Hayden Planetarium. This makes me so happy I want to cry. *Please*, I think, *as often as you can, put everyone on a bus and take them to the Museum of Natural History so they can see those short, hairy creatures planting their footsteps in the ash, or to the adjoining planetarium, where they can hear the Voice of Awe speak in hushed tones about how the universe and stars were born.*

Nor do I find that much has changed in the building where I attended junior high and high school. Given the disparaging remarks I have read on Facebook from people who used to live here, I expect to find dear old Liberty High crumbling to bits, gangs shooting it out in the parking lot. But the building seems no more broken down than before, with a new middle school on the hill and bulldozers breaking ground for a new addition. A guard checks that I have permission to enter, but I feel no more threatened than when I attended classes here in the seventies.

Beyond the entrance, the hall to the cafeteria is hung with photos of every senior class that ever graduated; by coincidence, my father's class photo lines up above my own. Along the walls, two large cases display our school's debate trophies, although the cases hold no trophies dated later than the seventies, as if Eric and I mark the dead end of an evolutionary branch that's gone extinct. The property taxes the hotels paid no longer support the schools, and after our coach died, no teacher seemed willing to give up evenings and weekends to supervise the team. Barry is buried somewhere in Florida, so my only way of mourning the first man I loved is to stand before these trophies, head bent, as if they are monuments to his grave. An added sadness comes when I notice the debaters atop the trophies, their miniature fingers raised to make a point, are all tiny brass-plated men.

The bell rings, and students flood the halls. I see the same scruffy kids who would have walked these corridors when I was a student, in roughly the same proportion of black to brown to white. The only difference is a few of the highest-achieving kids in each grade are gone, while most of the students who remain are poorer than the kids I went to school with, whose parents could always find work at the hotels. In a school this small, even a slight shift in demographics—the loss of the few most ambitious kids at the top, the addition of a few more disadvantaged kids at the bottom—can result in the perception that the school has gone from being a breeding ground for debate champions to a swamp of lazy illiterates whose failure to meet state-mandated standards for eighth-grade math proves the country we love is doomed.

In fact, when I return to the main office to keep my appointment with the principal, I discover that not one but two sections of calculus are being offered, with an equal number of physics sections and new electives such as Environmental Biology and Forensic Science. Although a black male student who enters Liberty High still has less than a fifty-fifty chance of graduating, the administration devotes far more attention to helping those black male students stay in school.

The principal, Jack Strassman, has a boyish face, kind eyes, and warm smile that make him far less intimidating than Mr. Van. The father of a daughter and a son, he is determined to convince students of both genders they can do whatever they take it in their heads to do, whether that means playing basketball or joining a ballet troop. Unfortunately, a lot of people in town are still sending the students who attend his high school other messages. The residents hold a fairly traditional view of female roles, he says, and the students themselves hold plenty of sexist attitudes, which govern their behavior not only in their academic pursuits but in their sexual and romantic relationships.

As was true when I went to school here, girls often rank among the top ten graduating seniors (my year, eight of the top ten were girls). When Strassman looks at the math and science classes, he sees a good mix of boys and girls. But in the last few years, the top scores on the math SATs went to boys. "I can think of boys going into math and boys going into science, but I can't really name a lot of girls."

As principal, Strassman finds himself the victim of a cycle in which the scarcity of students who go on in science or math creates a scarcity of well-trained teachers. "The last time I went looking for math teachers, it was a terrible experience. If I advertise for a social studies teacher, I could get a pack of applications up to here." He lifts his hand three feet above his desk. "But math . . . I'd be lucky if three or four apply. You can't blame them. Math majors can make so much money out there in the world." The pool of science teachers is slightly deeper, but he has trouble finding candidates with the ability to teach more than one subject—biology and earth science, say, or chemistry and physics. He found a replacement for his recently retired physics teacher by asking his biology teacher, Cindy Nolan, how she would like to teach the smartest kids, then sending her back to school to obtain her certification in physics. The students love her, Strassman says, especially the girls.

As if on cue, Cindy Nolan pokes her head in the door—she's here to escort me to her third-period physics class. Cindy is an exuberant blonde in a white shirt, dangly necklace, and black pants that even the most critical teenage girl would consider stylish. As she walks me to her room, she tells me that her father was a doctor who did Alzheimer's research, which inspired her fascination with biology, but she's more of a people person, so she decided to become a teacher.

"Hey, Mrs. Nolan!" the students sing as we enter the same room in which I studied physics. Rather than take my customary seat in front, I try to hide amid the rolls of duct tape, chunks of Styrofoam, and unused PCs in the back. In the cabinets, I see the same jumble of solenoids, generators, and dismantled radios that appeared to be antediluvian when my classmates and I used them in the seventies. Only the posters are new—the requisite photo of Albert Einstein sticking out his tongue, jokey slogans about physicists doing it with ENERGY, and advice to the effect that "Thirty years from now, it won't matter what shoes you wore, how your hair looked, or the jeans you bought. What will matter is what you learned and how you used it."

Thirteen rumpled boys sit slumped behind their desks, interspersed with five young women who clearly took more pains with their appearance. Earlier in the week, the students ran a race, and Cindy has them calculating each other's velocity and acceleration. Her jokes about her athletic misadventures—apparently, a video exists of Cindy bungee jumping from a bridge—create the sort of easygoing camaraderie that allows a teacher to humor her students into understanding that day's lesson. And yet the ratio of women to men is still significantly less than one to two. And this isn't an AP course. The lesson Cindy Nolan teaches is pretty much the same lesson on acceleration Mr. Yates taught us. When the students beg Cindy to slow down, I want to warn them their professors in college will recap this material in less time than it took the fastest student to run the twenty-meter race that provided data for their lab.

Then again, if you are going to encourage a third of the students at a small, rural high school to study physics, you can't gear the course to the top 1 or 2 percent. When the bell rings and the members of Cindy's next class, The Physics of Toys, come shuffling into the room, I am moved to realize that when I went to school, none of these kids would have been permitted to take a class with *physics* in the title. The boy in the wheelchair wouldn't have been allowed to attend school at all—in the old days, the building had no elevator or accessible bathrooms. Nearly half the students in Physics of Toys are girls. And I can't help but blink back tears when one of the three black guys shares a soul shake and a hug with one of the white guys, then one of the white girls, then the other two black guys. In my day, these kids would have been assigned to a stultifying course called General Science, if they hadn't dropped out by now. Instead, they are taking a class I wish I'd had a chance to take. Who wouldn't love designing and building a rocket that will carry an egg into space and bring it back via a parachute hidden inside its nose cone? The course doesn't require difficult calculations, but maybe if I had learned to design and build toys in high school, I wouldn't have been so clumsy in my chemistry and physics labs in college.

.   .   .

Lunch with Cindy Nolan and the chair of the Science Department, Mike Hazelnis, feels less like dinner with Wilmer Sipple and his wife than a meal with the coolest kids in school. With their training in science, they could be dining at a company cafeteria in Silicon Valley. Yet I doubt Mike would be any more excited discussing the latest app than describing the way he uses a SMART Board to animate the clouds and hurricanes when he teaches earth science. And Cindy tears up telling me about the boys who used to bring her hearts and lungs from the deer they hunted so their classmates could peer at these organs through a microscope and see all the miniscule trachea and bronchi.

When I tell the sad tale of my not being allowed to skip ahead in science and math (if nothing else, repeating the story has dulled my anger to the point where it no longer stings), they are appropriately shocked. The only criterion they use when selecting the eighth graders who will enter the accelerated track is "picking off the cream of the crop," although they do sometimes bend to parental pressure, which this year resulted in the inclusion of three girls who shouldn't be in the advanced classes "because they aren't open-minded thinkers." Most years, the advanced classes are half female and half male. Cindy isn't a scary person, Mike says, so girls are more willing to take her class. If we encourage them, if the guidance counselors advise them that they need the most challenging courses to get into college, then yes, the girls will sign up to take physics.

I venture my recollection that the faculty wasn't always this sensitive to the needs of their female students. "The previous head of the Science Department used to be much more male-centric," Mike admits. "I remember when we hired Cindy, he said, 'Oh well, there goes the department.'" Now, two of the six members of the science faculty are women. In the past, Mike says, "mainly the males did well in science in college." More recently, a female student named Carrie did a very impressive research project related to dark matter before she went on to Yale to study physics.

"Really?" I say, torn between relief that it is now routine for girls from my hometown to go on to Yale to study physics and my disappointment that female physics majors from my alma mater are a dime a dozen.

Mike is right that the SMART Board makes his lesson on contour maps more exciting than Wacky Wach's lecture on the same subject (although filmstrips must have seemed an exciting improvement over chalkboards). But the students in fifth-period earth science need so long to grasp the essentials, I spend as much time fidgeting as I used to, the only difference being that some genius has put tennis balls on the bottoms of the chairs to keep the legs from squealing.

For all that, I would rather spend the rest of the day in Mike's earth science class than meet the female physics students Cindy has arranged for me to interview. I ought to be overjoyed that even in my hometown, three young women are willing to proclaim their interest in science and math, but the prospect of spending time with teenage girls who remind me of my former antagonists at dear old Liberty High triggers so much anxiety I conduct the worst interview of my career, not managing to attach names to the comments in my notes, let alone descriptions of what each girl looks like, other than that all three are attractive and fit and two of the three are blondes.

One of the blondes describes herself as "on the girlie side" and professes to enjoy the shock on people's faces when she tells them she wants to become a high school science teacher. The other blonde tells me her math teachers in the lower grades were terrible, but she loved being good at solving problems, and she still loves how logical the subject is. She "pretty much knows" she will go on to teach math and maybe physics. She already has gotten some experience tutoring, and she loves when people come back to her and say, "Yeah, thanks, I got such good grades because of you."

The gang's leader, let's call her Amanda, says she became interested in science because her mother is a math teacher who "gets so into math problems it's ridiculous." Amanda and her friends used to spend Friday nights with Amanda's mom because she made doing math so much fun. "She'd always tell me cool things like 'a line is never a straight line in outer space because the space out there curves.'" But Amanda's interests are more down-to-earth. "I'm going to be a surgeon. Thoracic, or abdominal." When Ms. Nolan had her biology classes examine those deer organs, Amanda was "all into it, holding the heart and lungs, looking at the anatomy. I'm the kind of person, when someone sneezes, I tell them, 'I can fix you! I can fix you!'"

I marvel that my alma mater has been transformed into a feminist heaven. "The cool kids are now the smart kids," Amanda says. The smart kids are the ones you see dancing at homecoming. A girl can wear a T-shirt that proclaims I LOVE MATH without sabotaging her social status. In their cluster of friends, *dork* is a term of endearment. No divisions exist between the kids who are good at academics and the kids who are good at sports. (If she doesn't practice with her soccer team, Amanda says, she gets "really antsy and paces around." Athletics provide her with an outlet for her frustrations that I didn't have.) "All of us are into art," Amanda says. "We're all in some kind of music group."

I ask why, if girls have become so tolerant, I hear so many horror stories from my friends whose daughters have been objects of derision by their classmates. Amanda screws up her face and says, "Yeah, girls are still mean. It's retarded. Most of my friends are guys. The girls on the soccer team are one big family. And we hang out a lot with the guys on the boys' team." She is especially fed up with girls who aren't "logically dominated." Her friends chime in on this one. "The other girls just don't think logically about how they act. It's not logical to lie in a relationship! Then why lie? Does that make any sense?"

I have to agree that lying in a relationship makes very little sense. But the rest of what Amanda says doesn't compute. The burden these three young women carry—not only to be smart, but also to be attractive, good in music, art, and sports—strikes me as overwhelming. The ratio of boys to girls in Cindy Nolan's physics classes still heavily favors men. The teaching of science and math in elementary school still strikes these girls as "terrible." They still perceive people as "shocked" at hearing that a pretty girl might want to teach science or math. None of the three intends to go on for a PhD. And even with a (general, non-AP) course in calculus and Cindy Nolan's class in physics, they are in for a rude surprise when they take introductory physics in college. I consider pointing this out, but the girls already regard me with the scorn many young women reserve for old-timey feminists who see sexism where there is none, the way I regarded my parents as paranoid for thinking anti-Semites lurked around every corner.

Only when I ask where the girls want to go to college do they think I have any wisdom to offer. Amanda is afraid the really good universities won't accept an applicant from a crappy school like Liberty. All three girls suddenly seem so nervous I feel sorry for them. Growing up in a world where no one expected anything from a girl, I never experienced the pressure to fulfill anyone's expectations.

"Where did *you* go to college?" Amanda asks, and when I tell her Yale, her eyes grow round as petri dishes. "Yale! You went to *Yale?*" She turns to the other girls. "She went to Yale!" Then she turns back to me. "Yale took someone from Liberty? What was your secret?"

The word *secret* makes me wince. Then again, Amanda's passion to become a thoracic surgeon seems as genuine as my passion to become a physicist. Thirty-five years ago, far fewer women were applying to Yale. If I had been under as much pressure as Amanda to get into an Ivy League school, I might have tried gaming the system, too.

She is still looking at me expectantly. *Be yourself,* I want to say. *When you write your admissions essay, describe the excitement you felt dissecting deer hearts in Cindy Nolan's biology class. The problem isn't getting into a good university; it's adjusting to the hours of brain-cracking homework once you're there. The problem isn't getting into a decent medical school; it's that even a woman as ambitious as you might be tempted to choose a career that requires fewer years of training and allows its practitioners more reasonable hours than being a thoracic surgeon, especially if you want to marry and have a child.* I want to tell her how hard my roommate Laurel worked to get into medical school and train as a trauma surgeon, how much Laurel needed to give up to accomplish what she accomplished.

The bell rings, and I ask Amanda if anything might keep her from achieving her dream. She scowls, as affronted as I would have been if someone had suggested I didn't have the stamina or intelligence to become a physicist. "Yeah," she says, "I might not become a surgeon if someone cut off my hands."

．　．　．

I have forgotten how tiring a school day is. By the time Amanda and her classmates have taken their seats for their eighth-period physics class, I can barely keep from laying my head on my desk. Cindy does a great job teaching the standard lesson about all the different forces that act on a concrete box sitting on a floor. But I remember listening to this same lecture and wondering what was wrong with me for thinking the explanations went only so far. What is gravity? I wanted to ask. *Well, gravity is the force that pulls an object toward the earth.* But what is the force that pulls objects toward the earth? *I just told you! The force is gravity! See? It's represented by this arrow pointing down!* I didn't care how many seconds it took a bowling ball to roll down a ramp. I wondered why the earth tugged bowling balls down ramps at all. But these were the questions I couldn't ask—not unless I wanted my classmates to roll their eyes and mutter *brown noser, teacher's pet.* This was years before the great wave of popular science writing. Even after I got my driver's license and was able to travel to the community college, I found little but a biography of Einstein and a few frail monographs by German physicists with dots over the vowels in their names. I marveled at Erwin Schrödinger's description of his cat, which, due to the probabilistic nature of quantum decay, could be alive and dead at the same time, and Sir Arthur Eddington's revelation that the nucleus and electrons in an atom are so far apart, it's a wonder concrete boxes—and high school physics teachers and their students—don't go plummeting through the classroom floor. But every page I read raised more mysteries than it resolved.

When the bell rings, Cindy is so swamped with questions I mime my gratitude and slip out. I'm exhausted, but I stop to talk to the calculus teacher, Mr. Olson. Of everyone in the school, Mr. Olson is the only faculty member who dates back to my time—he entered the system just as I was leaving. He is also the only teacher from whom I get the same disapproving vibes I used to get from Ed Wolff. It's as if some radioactive fallout from the bad old days clings to his lapels. When I ask if he has any insights as to why fewer girls than boys go on in math, he bristles and says he resists the notion that any such disparity still exists. There's no longer any stigma to girls doing well in math, he says. The girls in his classes experience no more anxiety than the boys. If anything, the girls are more mature and more willing to do the homework. At least 50 percent of the math teachers in his department are women, and when the male teachers retire, they probably will be replaced by females who, Mr. Olson claims, don't challenge the administration the way their male counterparts do.

The longer we talk, the more I get the sense that Mr. O. thinks women have an easier time studying math, and an easier time finding jobs. I can forgive him for thinking the girls in his classes don't suffer any more stigma or anxiety than the boys—the girls in Cindy Nolan's class told me the same thing. But it's possible that a significant number of girls don't make it as far as Mr. Olson's calculus class (or Cindy Nolan's physics class) because the stigma and anxiety they suffer in elementary school and junior high prevent them from signing up. I can hardly take umbrage at Mr. Olson's supposition that the principal finds female teachers—Strassman admitted to me earlier he prefers working with female members of student government because the girls ask his permission before they institute a change, while the boys simply go ahead and do what they want to do. But the principal also said he tries to hire female math and science teachers because he considers it vital to provide the female students with much-needed role models. The reason it's easier for women to find jobs teaching math is that so many schools are under so much pressure to hire those few women who have the credentials.

I ask Mr. Olson why, if female students have such an easy time majoring in math and science, there are so few women in those careers, and he admits his son, who is majoring in aeronautical engineering at Clarkson, has very few female classmates, a circumstance Mr. Olson dismisses as unavoidable because "guys are more hard-wired to build things" and "better able to relate to stuff they've done with their dads." I consider pointing out that biologists have yet to find a gene for "building things" and that if dads invited their daughters to help them build go-carts, forts, dollhouses, radios, rockets, and model airplanes, the daughters might be able to relate to engineering, too. But I am in no mood to provoke an argument, so I tender an abrupt good-bye and leave.

CHAPTER THIRTEEN

Integration and Differentiation

One of my motives for returning to my hometown is that I am determined to hear my nemesis, Ed Wolff, say, "Eileen, I'm sorry I didn't recognize how talented you were in math." And I want to render a belated thanks to my trigonometry teacher, James Gallagher, for helping me to study calculus. Senior year, he loaned me the textbook he used in college. When I got stuck, he provided guidance, although he sometimes said he didn't know any more than I did. I have no idea if Ed Wolff was aware of what his friend was doing, but I felt like a child whose mother sneaks her dinner without her father knowing.

The dry-as-dust calculus book my teacher lent me was written in the fifties by an MIT professor named George B. Thomas Jr. Paging through that same book now, I realize what a huge role textbooks play in determining who does or doesn't appreciate the beauties and joys of math or, for that matter, make it to the end of the semester. To get some feel for this phenomenon, one need only peruse *Math Doesn't Suck: How to Survive Middle School Math without Losing Your Mind or Breaking a Nail*, by Danica McKellar (before McKellar earned her bachelor's in math from UCLA, she starred as Winnie Cooper on *The Wonder Years*, a show I particularly enjoyed because Winnie and her boyfriend, Kevin, were born the same year I was). With its word problems that involve not baseball statistics and rockets but boyfriends, beads, and Barbies, McKellar's book (along with its sequels, *Kiss My Math: Showing Pre-Algebra Who's Boss; Hot X: Algebra Exposed;* and *Girls Get Curves: Geometry Takes Shape*) probably has done more to encourage girls to excel in math than any Harvard task force might accomplish. Unlike the girls in McKellar's audience, I didn't need a television star to convince me math doesn't suck. But I would have given anything to hear that same star assure me only 9 percent of the boys she polled said they like girls who dumb themselves down, while 52 percent said they hate when girls pretend to be dumb (for the record, the other 39 percent don't notice or care).

To appreciate what McKellar is doing, compare her books to their counterparts in the Dummies series. When the authors of the Dummies books in geometry and trigonometry endeavor to excite their readers about the careers those subjects might enable them to pursue, the list is composed of fighter pilots, quarterbacks, sailors, bridge builders, grass-seed sellers, architects, and surveyors.[1] Nearly all the humor derives from the spouting of mathematical theorems by male football players, pirates, gangsters, baseball coaches, golfers, and grotesquely nerdy scientists who use their prowess in math to scare away attractive female dates. The only women are a ring of tennis players engaged in a round-robin tournament, a woman making a *pie* iced with the value of *pi*, and a damsel in distress awaiting rescue by a knight mathematically gifted enough to calculate the correct placement of the ladder. All the sidebars about famous mathematicians focus on men, except for a reference to Hypatia, who ended up

---

1. In *Trigonometry for Dummies*, the author asks her readers to suppose that two surveyors are taking measurements for a road-paving crew, a request that whisks me back to an afternoon when I learned a local surveying company was looking to hire students and I walked in and asked the owner if he would hire me.

"Are you crazy?" he said. "How could we send a woman out in the field with a surveying crew?"

"Why?" I said. "What would happen?"

"If you don't know," he shouted, "ask your father!" then waved me out the door. I wasn't sure if he meant the men on the crew might rape me, or my presence would prevent them from telling dirty jokes. But I didn't bother to ask my dad, knowing that whatever objections the owner of the company might entertain, my father would share them.

being torn to bits by a mob. Not a single black, Asian, or Hispanic person of either gender makes an appearance. And the only Native Americans figure in a joke about squaws, hippopotamus hides, and papooses (resulting in the punch line that "the squaw on the hippopotamus is equal to the sons of the squaws on the other two hides") and a hokey "legend" about the Indian brave SohCahToa, whose name is offered as a device to help students remember how to calculate the sine, cosine, or tangent of any angle.

Maybe I fell in love with calculus because, unlike geometry and trigonometry, it didn't seem to be designed for men. Not that calculus has no use—without calculus, we wouldn't be able to predict the behavior of any object that isn't static. But the derivations are far more abstract than in geometry or trigonometry. Studying calculus, you find yourself marveling, *How did anyone think of that?* Yet the beauty is so apparent you feel you could have invented calculus yourself, if only you were a genius. It's like reading a poem that expresses exactly the way you feel and thinking you could have found the words yourself, if only you were Emily Dickinson.

Or maybe calculus is more like magic. You add an infinite number of infinitely tiny quantities and end up with a concrete sum. Imagine dreaming you have been to Oz, only to wake the next morning to find a ruby red slipper on your pillow.

When I drive up to Ed Wolff's house, he steps out on his porch and waves. He must be in his seventies, but he still seems larger than life, and not only because of his huge size and red flannel shirt. "You've barely changed," I gush, and Ed booms, "Neither have you!"—both of us lying, both of us telling the truth.

Inside, Jim Gallagher sits quietly in the kitchen. He seems not to have aged so much as grown thinner and more brittle. I remember him telling me he had just gotten over a serious illness and am stricken with dread that the last thing he needs is to waste time helping me to heal a psychic wound I suffered in seventh grade.

Ed motions me to sit in one of the two giant recliners, then sinks into the other and puts up his enormous, sneakered feet. I ask about his past, and he talks about his difficult childhood, his early struggles as a teacher, and how he came to love his job. I can't remember him as anything but low-key and sardonic. But I am starting to think I understood nothing about the man. For one thing, he isn't that good in math. He always liked geometry because its logic appealed to him. "But I had trouble with calculus in college," he says. "I sure did."

Unlike Ed, Jim enjoyed calculus, precisely because it was so challenging. He was sorry when the number of qualified students who could take calculus grew too small and the administration discontinued the course.

I ask if it was fair to deprive those few qualified students the opportunity to take a class they would need in college, and Ed booms that in his personal opinion, "there are certain kids who simply can't do certain things, whether English or math. Not every kid can go to college. That's a crock. You probably find it too, Eileen, don't you, that you have students who can't read or write and need remedial English." (In fact, even the kids who enter the University of Michigan with weak backgrounds in reading and writing often manage, with the aid of remedial courses, to complete their degrees.) These days, Ed says, 40–50 percent of the students in Liberty can't meet the minimum standards for eighth-grade math. Besides, the state recommends only the top 2 percent of any class be expected to take the AP exam in calculus, and in Liberty, that would mean only two or three kids per year. With the money crunch, a school can't offer a class for two or three kids.

Well, if 98 percent of American high school students are unfit to take an advanced course in math, aren't we doing something wrong?

Maybe, Ed says. Then again, a lot of people have trouble with math because of their attitudes. They think they can't do it. They're scared of it.

That's right, Jim chimes from the kitchen. One downside to teaching math is telling people what you do for a living. Every time he sees a doctor, the doctor asks him what he does, and when Jim tells him, the doctor says how much he hated math, which leads me to wonder how many doctors Jim has been seeing, and for what.[2]

---

2. The illness must have been serious; I was heartbroken to hear that he died not long after.

It's true, Ed says. Most elementary school teachers had trouble with math. And the parents can't help the kids. Or they don't care. Still, he says, you can't just ignore the students who don't know what's going on. Maybe if teachers didn't have to worry about their kids passing the Regents Exam, they could cover more advanced topics, or dig more deeply into the topics they do cover.

All this talk about who is or is not entitled to take advanced math classes makes me squirm. I begin edging toward the question I have come to ask. "So," I say, "did either of you see a difference between the math abilities of boys and girls?"

In chorus, Ed in his boomy baritone and Jim in his silvery tenor, sing out: "The girls were better!" Without a doubt, Ed says. In college, the girls in Ed's study groups had a much better idea of what was going on than he did. It comes back to him that two of my female classmates did so well he asked them to help teach remedial math in summer school. (I remember how hurt I was that Ed asked Sharon and Nancy to work for him but didn't ask me. Then again, why would he want to work with someone who gave him such a hard time?) Ed says that before I drove up, he went through my yearbook, and the photos jogged his memory as to the number of girls who were good math students. Nancy, for one. And Sharon. And didn't my classmate Janet go on to major in math at Barnard? He could swear Janet reported back to him that she scored so high on the math placement exam she was assigned to advanced calculus, despite not having taken intro calculus in high school.

I ask what Ed remembers about my presence in his classes. He shrugs, as if to dismiss my bad behavior. "Oh, go ahead," I say. "I was a pain in the ass." And then, unbelievably, I hear myself apologize. "I'm a teacher now. I know what it's like to have a student give you a hard time. I'm really sorry."

"Aw," he says, "forget it," and with one plate-sized hand waves off my apology.

I ask Ed to name the best math student he ever taught. We laugh about a neighbor of mine who went on to achieve fame as Bernie Madoff's accountant. "I looked him up in the yearbook," Ed says. "From what I remember, he was a pretty smart guy."

But the smartest kid? That had to be David G. "David used to come up from the elementary school by bus, and he would sit in the main office and wait for me to come and take him back to my office so he could play around with whatever calculations I would give him. So one day, he was waiting for me to come out, and the secretary in the office says to him, you know, the way you do to a kid, 'You must be really smart,' and David says, 'Yeah, I am!'" Ed laughs. "David used to drive Harry Wach crazy. David was so bored he would sit there right in Harry's class doing math, and Harry wanted to demand that he pay attention. I warned him not to. I said, 'Harry, you'll be sorry. Just let the kid alone.' And I was right. Harry made David put away the math, and David started paying attention, started asking questions Harry couldn't answer and catching him in mistakes. I said, 'I told you to let him be!'"

"Hmm," I say, wondering if I should point out that's how I felt in Ed's class. I ask if he knew I majored in physics and math at Yale.

No, Ed says, he had no idea where I went or what I did. Before I showed up, he called my social studies teacher, Mr. Heffley, and asked why I might be coming to interview him, and Mr. Heffley said, "Oh, don't worry about her, she went on to do something in English, she became a writer."

Ed walks me to his porch and I drive off, smacking my head in disbelief that after all these years, I ended up apologizing to him. Why couldn't he have greeted the news I had majored in physics at Yale by saying, "What an accomplishment!" Even more infuriating, why did he laugh at David G. for giving Harry Wach a hard time, yet chide me for giving him a hard time for the same reasons? I wonder if he knows that after David G. earned his physics degree from MIT, or maybe it was Harvard, he came back to Liberty, moved into his parents' basement, and spent the next few decades doing little more than screwing around with computers, although I recently heard he got a job in the next town teaching math.

CHAPTER FOURTEEN

The Women Who
Don't Give a Crap

For the next few months, I read everything I can find about the reasons women do or do not pursue careers in science. I ask everyone I know why she did or did not major in physics or math in college, what obstacles she encountered when applying for jobs or coming up for tenure. Everyone has a story to tell, even the women who stopped taking science and math in high school.

Especially the women who stopped taking science and math in high school.

"One reason I didn't go on was because in high school I was so good in all of my subjects, and math and science seemed harder than everything else, so I thought I wasn't good at them," says Jamie Saville, the administrator who helps run the program for Women in Science and Engineering (WISE) at the University of Michigan. "I didn't realize science and math are harder for most people. They're just harder subjects. In high school I took a computer programming class and loved it. But I was one of only two girls in the class, and I always felt uncomfortable. I took another programming class in college and had a similar experience, so I never took another computer course again."

Cinda-Sue Davis, who directs WISE, tells me that her daughter was asked to fill out her schedule for her junior year in high school. The guidance counselor gave her a list of possibilities, but Davis's daughter had already taken some of the courses, and the rest didn't look interesting. She checked the website to see if any other classes were meeting at the right time, and found an engineering technology course her older brother liked. But the guidance counselor never mentioned that as an option.

This is far from the only example of female students receiving faulty advice about which courses to take. Soon after my meeting with Ed Wolff, I contact the classmate he mentioned, Janet, to find out if he is right about her going on to major in math at Columbia. When she arrived at Barnard, Janet writes me in an e-mail, she had to take a math placement exam. She tested so high they placed her into an advanced calculus class. But the professor had a thick French accent, and she had no prior experience with the subject, so she ended up with a B- in the class. No one suggested she take a regular calculus class instead. "It was the end of my undergraduate career at Barnard," Janet writes, "and the end (at that point) of my lifelong quest to study Astrophysics and discover life in space."

And what of Carrie, the young woman Cindy Nolan mentioned who grew up in Liberty twenty years after I did and went off to study physics at Yale? When I find Carrie via Facebook, I discover our childhoods followed eerily parallel lines—literally, given that the street Carrie lived on ran parallel to my own. As a kid, Carrie enjoyed writing, but she also loved to watch the chicks hatch in the incubator at the back of the classroom and read books about the solar system. She loved that so many scientific questions remain unanswered. When she studied English and history, she thought, "Okay, this is what it's like, this is what the subject is. But in math and science, it was obvious we were only getting the tip of the iceberg."

As a junior, Carrie became fascinated with the question of what sort of dark, invisible matter might account for the excess gravitational pull that scientists can detect disturbing the objects we see. Her teacher took the class on a field trip to a library at SUNY Albany, and Carrie studied other people's research. She knew she wasn't doing the real thing herself, but she glimpsed how exciting that might be and dreamed of solving the mysteries of dark matter.

Like me, Carrie loved calculus. "Everything in math had been so tangible until then. Then we got to limits, and I thought, 'Wow, this goes way beyond counting pennies.'" Carrie found Mr. Olson to be more "teachery and dominating" than Cindy Nolan, but he was "very relatable, kind, and generous," and she never felt any vibes that she couldn't do math. Everyone encouraged her, especially her parents. "That gave me a lot of confidence. And believe me, I needed that confidence when I got to Yale."

That first semester, Carrie found the lectures in physics "just so over my head." She listened to the professor tell stories about race-car driving to illustrate some concept, "and I kept waiting for what he was saying to link to the problem sets I needed to go home and do. But the stories never did connect to anything. I was being asked to teach myself the class. Nothing we did in high school prepared me for that. It caused me not a little stress to look at the problem sets and have no idea how to do the problems."

Carrie was more fortunate than I was. By the time she got to Yale, the Physics Department provided undergraduates with a tutor. Two days before each problem set was due, she would visit a Russian doctoral student who led her through the nuts and bolts of that week's problems. But she couldn't get over her shock that other students could do the work while she was having so much trouble. She managed to get a B+ but abandoned the idea of a degree in physics. "Maybe if I had been better at it, I would have enjoyed it more. Also, I was learning that the world of academic pursuit was wider than I had thought. I would hear about courses in linguistics, or anthropology, or women's studies, and I didn't even know how to choose. I had thought English was just reading a book and writing a report on it. I didn't realize history could disrupt the national narrative."

By sophomore year, Carrie had started an internship as a Dwight Hall Urban Fellow in planning and development. The work struck her as more likely to decrease the suffering of everyday people and bring about a more equal society in ways science couldn't. She also started to see that "gender is so important, that gender influences a lot of decisions in a person's life." After graduating, she attended Yale Law School, where she pursued her newfound passions—public interest law, poverty law, civil rights litigation. She admires her friends who went on in science, but she sees them as engaged in an "isolating, solo enterprise. I think we pick our paths based on what activities we find rewarding, what we get the most praise for. But I do also think some of that is related to gender. A person who wants to be more nurturing, more self-sacrificing, more helpful, well, physics might not be the best choice for that person."

For all that, Carrie remains "really interested in cosmology." Her fiancé jokes that when they are old, they will hire tutors in math and science. And Carrie vows that one day she will read all of Feynman's lectures, not just the few she read in high school, in the book one of her teachers gave her as a gift. Those were the lectures that inspired her to study physics in the first place.

Whether women like Carrie, Janet, or me drop out of programs in science and math because we are innately incapable of succeeding at the same level as our male peers or because the culture in which we live discourages or hinders us has long been a matter of study. In 1980, a large group of American middle schoolers was given the SAT exam in math; among those who scored higher than 700, boys outperformed girls by thirteen to one. And yet, scoring 700 or higher on the SATs, even in middle school, doesn't necessarily reveal true mathematical creativity or facility with higher-level concepts. In 2008, the American Mathematical Society published data from a number of prestigious international math competitions to track the "one-in-a-million" standout performers. The American competitors were almost always the children of immigrants and only very rarely female, but this wasn't the case in other countries: between 1959 and 2008, tiny Bulgaria sent twenty-one girls to the International Mathematical Olympiad, while the United States, from 1974—when it first started entering the competition—to 2008, sent only three girls (no woman made the American team until 1998). The AMS study went much further in capturing performance at the genius end of the spectrum than did the study of those middle schoolers taking the SATs; it also measured the performance of girls in other countries. The study's conclusion? The scarcity of women at the very highest echelons "is due, in significant part, to changeable factors that vary with time, country, and ethnic group. First and foremost, some countries identify and nurture females with very high ability in mathematics at a much higher frequency than do others."

In 2010, the American Association of University Women published a report called *Why So Few? Women in Science, Technology, Engineering, and Mathematics,* whose authors argue it's illogical to attribute the scarcity of women to innate biological differences because the number of girls earning very high scores on math tests has gone up so quickly in recent years. Remember those thirteen-year-olds who scored 700 or higher on the math SATs? Thirty years later, the ratio of boys to girls had dropped to three to one. That's still a significant disparity, but if girls were constrained by their biology to perform poorly, how could their scores have risen so dramatically in such a short time?

In elementary school, girls and boys perform equally well in math and science. Only in junior high and high school, when those subjects begin to seem more difficult and girls become more conscious than boys of their social status, do the numbers diverge. Although the percentage of girls among all high school students taking physics rose from 39 percent in 1987 to 47 percent in 1997, that figure has remained constant into the new millennium. (Even in Cindy Nolan's class, I counted only five girls to thirteen boys, and that was an introductory, non-AP course.) The numbers become more alarming when you look at AP classes rather than general-level physics. At the AP level, only 41 percent of the seats in Physics B and fewer than a third in Physics C are occupied by girls. Worse: only half the girls who take the Physics B course take the AP exam (as compared to 65 percent of the boys), while 61 percent of the girls who finish Physics C take the test (compared to 78 percent of the boys). Still worse: only half the girls who take the Physics B exam receive a passing score (compared to 62 percent of their male classmates), and only 60 percent pass the Physics C exam (compared to 72 percent of the boys). The statistics tend to be better in AP math, but far worse in computer science. Maybe boys innately care more about physics and computer science, or boys are better at taking exams. But an equally plausible explanation is that boys are conditioned to tough out difficult courses in unpopular subjects, while girls, no matter how smart, grow weary of their classmates teasing them about being the only girl in the room and receive fewer arguments from their parents, teachers, or guidance counselors if they drop an advanced physics class or shrug off an AP exam.

Londa Schiebinger, a professor at Stanford and the author of *Has Feminism Changed Science?*, points out that even though girls and boys start out in relatively equal numbers in science classes, performing at roughly the same level, it takes four hundred ninth-grade boys—but two thousand ninth-grade girls—to get one PhD scientist. Part of the problem, Schiebinger writes, is that girls are raised to be modest, while boys learn to exaggerate their intelligence, their success, their prospects in life, even their height. Girls who have been trained to underestimate their talents encounter boys who overestimate their talents; the girls take the boys' estimations of their skills at face value and think even worse of themselves. According to a study of undergraduates cited in Schiebinger's book, three-quarters of the women who gave up science, compared with fewer than half the men, mentioned low self-esteem as their reason.

That the impressions a student picks up from her culture can affect her ability to perform on an exam has long been known. In a 1999 study, a sample of University of Michigan undergraduates with equally strong backgrounds and abilities in math were divided into two groups. In the first, the students were told men perform better on math tests than women; in the second, the students were assured that despite what they might have heard, there was no difference between male and female performance. Both groups were given a math test. In the first, the men outscored the women by twenty points; in the second, the men scored only two points higher.

Before Shelley Correll became a sociology professor at Stanford, she taught high school. No matter how poorly the boys in her chemistry classes performed, she noticed they continued to think of themselves as good at chemistry, while no matter how well the girls performed, she couldn't convince them they had any sort of scientific ability. Later, Correll devoted her career to studying this phenomenon, concluding that "boys do not pursue mathematical activities at a higher rate than girls do because they are better at mathematics. They do so because they think they are better."

Then again, not many Americans of either gender are dying to become scientists. A study of thirty-nine male and forty female Yale undergraduates revealed that both groups held negative attitudes toward math and science as opposed to languages and the arts (although the women showed more negative levels than the men). In his angry and insightful book *Nerds: How Dorks, Dweebs, Techies, and Trekkies Can Save America and Why They Might Be Our Last Hope*, David Anderegg, a child therapist and psychology professor at Bennington College, makes a convincing case that anti-intellectualism runs deep in American culture. As a child therapist, Anderegg daily sees the pain caused by the ostracism of his young clients, who suffer their classmates' taunts as early as kindergarten, if only for doing too well in class. "I still talk to kids every day who tell me, 'Yeah, I could be getting As in school instead of Cs, but I don't want people to think I'm a *nerd*,'" Anderegg writes. "I still talk to kids who want to die because other kids scream at them, literally scream at them, in the halls of junior high schools, 'get away from me, nerd!' because they did really well on a math test." A child's own parents sometimes are the ones to convey alarm that he or she is projecting a nerdy demeanor.

Although the teasing and bullying diminish as children grow up, Anderegg argues the stereotypes picked up in the schoolyard ("we all know one cannot be both sexy and smart") can persist into adulthood. "Kids in seventh grade think nerds are just awful, and they make life decisions accordingly. Some of those life decisions involve not studying science and math, which is a bad thing for them and for our country."

Anderegg fears that America's escalating obsession with physical beauty is making science and math as unappealing for men as it always has been for women. But the emphasis in his book is on the ostracism faced by male nerds and geeks; in the world he portrays, women are far more likely to be the object of sexual longing by male nerds and geeks than to be the nerds or the geeks themselves (the paradigm for this dynamic being the truly awful reality show *Beauty and the Geek*). This is a view supported by the AMS, whose study found that native-born Americans of both genders steer clear of math clubs and competitions because "only Asians and nerds" would voluntarily do math (an attitude as troubling for its anti-Asian racism as for its antiscience bias). "In other words," the authors write, "it is deemed uncool within the social context of USA middle and high schools to do mathematics for fun; doing so can lead to social ostracism." And yet, the effects of America's antinerd bias is even worse for females than males. "Consequently, gifted girls, even more so than boys, usually camouflage their mathematical talent to fit in well with their peers."

From 1974, when the United States began competing in the International Mathematical Olympiad, until 1998, when Melanie Wood, a cheerleader and math prodigy from a public high school in Indiana, made the team, all members of the American squad were male. Wood, who went on to study for her doctorate in math at Princeton, told a reporter from the *New York Times*, "There's just a stigma in this country about math being really hard and feared, and people who do it being strange. . . . It's particularly hard for girls, especially at the ages when people start doing competitions. If you look at schools, there is often a social group of nerdy boys. There's that image of what it is to be a nerdy boy in mathematics. It's still in some way socially unacceptable for boys, but at least it's a position and it's clearly defined."

If anyone needs proof of the stereotypes that continue to shape women's psyches, he or she need only watch the popular television show *The Big Bang Theory*. The show portrays the lives of an awkward but endearing Caltech physicist named Leonard; his roommate, Sheldon, a childlike genius who has trouble interpreting the behaviors of his fellow earthlings; Howard, a hunchbacked Jewish aerospace engineer who, at the start of the show, lives with his overbearing, infantilizing mother; Raj, a pudgy particle physicist who loses his voice when a woman enters the room; and Leonard and Sheldon's neighbor, Penny, an attractive blonde who has moved to LA to make it as an actress. *Wikipedia* sums up the show's dynamic: "The geekiness and intellect of the four guys are contrasted for comic effect with Penny's social skills and common sense."

Although several of the scientists on the show are women, Howard's girlfriend (and later his wife), Bernadette, speaks in a voice so shrill it could shatter a test tube; Sheldon's "girl friend who is not a girlfriend," Amy, is played by Mayim Bialik, an actress who really does have a PhD in neuroscience but is in no way the hideously dumpy woman she portrays on *The Big Bang Theory*. I admit I often lose ten minutes laughing at the show, which is witty and well acted. And an argument could be made that Penny, through her amused and friendly tolerance of her neighbors' foibles, encourages the viewer to appreciate their eccentricities and passion for science.

But even with Penny's—or the viewer's—tolerance, the scientists on *The Big Bang Theory* are portrayed as alien. And who wants to be a mascot, however charming? What remotely normal young person of either gender would want to enter a field populated by misfits like Sheldon, Howard, and Raj? And what remotely normal young woman would want to imagine herself as dowdy, clueless Amy rather than stylish, buxom, math-and-science illiterate Penny?

Although Americans take for granted that scientists and mathematicians are male geeks, this isn't necessarily true of other cultures. A former Romanian math champion who went on to earn a degree at Harvard told the *New York Times* that in her country, math isn't only about being a nerd. "It's about having intuition. It's about being creative." Germany, Canada, Japan, Switzerland, Norway, and Korea rate the same as—or worse than—we do in producing female physicists, but Italy, the former Soviet Union, and Portugal can boast far better numbers. Meg Urry told me the faculty at the space telescope institute where she used to work is very international. "The women from Italy and France dress very well—what Americans would call sexy. You'll see a French woman in a short skirt and fishnets, that's normal for them. The men in those countries seem able to keep someone's sexual identity separate from her scientific identity. American men can't appreciate a woman as a woman *and* as a scientist; it's one or the other." She definitely knows female American scientists who play down their looks. "This one woman, the men would come on to her every time she stepped out of her room, so she cut her hair, she started wearing jeans and baggy shirts. 'I'm not that,' she was saying, 'take me seriously as a scientist.'"

Rather than attribute American women's reluctance to major in science to the culture in which they are raised, Urry thinks most of her colleagues believe women simply aren't as interested as men in the physical sciences, and the men don't see this as a problem. "They think that if someone hasn't been interested in physics since birth, they can't be any good. But studies show that an early interest in science doesn't correlate with ability. You can be a science nut from infancy and not grow up to be good at research, or you can come to science very late and turn out to be a whiz."

Meg wishes she could convince her colleagues that people who drop out of physics tend to be as smart as the people who stay. "We race through the material in the intro courses as if we were going through a dose of salts." She thinks offering some kind of boot camp for entering students would be a great idea. "There have been conversations about that, but there's this worry that such a program might stigmatize the students who take part in it. Personally, I would tell them to take a summer course in physics and calculus before they come." She shakes her head. "The trouble is, most students don't know how far behind they are until they get here."

Physicists and mathematicians tend to believe that geniuses know they are geniuses from an early age and couldn't possibly be discouraged from rising to the top. And yet, the most gifted young women might experience even more frustration than their less talented peers at not being allowed to develop their talents, which might make them more likely to drop out earlier. And girls who are gifted at science and math might see more options than boys. The researchers who studied the winners of those international math competitions also looked at American students who compete for Davidson Fellowships, which are awarded to prodigies in literature, philosophy, and music, as well as science and math. "Profoundly gifted children are frequently multitalented," the study found. When deciding which talent to pursue, "they usually invest more of their effort in the fields that provide more positive feedback."[1] The authors of the study mince no words in putting to rest Summers's hypothesis that a gender-based disparity in aptitude at the prodigy end of the spectrum might explain the scarcity of tenured female mathematicians. "There exist many girls with profound intrinsic aptitude for mathematics; however, they are rarely identified due to socio-cultural, educational, or other environmental factors."

It's even possible that gifts in science and math aren't identifiable by scores on SATs or intensely competitive and stressful international competitions. Of all the statistics cited in *Why So Few?*, the discovery that really made my eyes pop was that fewer than one-third of the college-educated white males who populate the ranks of American engineering, computer science, math, and the physical sciences scored higher than 650 on their math SATs, and that more than one-third scored below 550. In the middle ranks, hard work, determination, and encouragement might be as important as raw talent. But even at the very highest end, genius might be more a matter of confidence and intuition than scoring in the top percentile on an intelligence exam.

According to Richard Feynman's biographer, the physicist's IQ was "a merely respectable" 125. What made him a genius was his ability to visualize the phenomena he was studying, his insistence on inserting himself into the rhythm of the dance of particles. What made him one of the greatest physicists of the twentieth century was his confidence that if he found a question interesting, the answer was

---

1. This finding is substantiated by Catherine Riegle-Crumb, a researcher at the University of Texas at Austin who wrote to tell me that "gender differences in high school math and science achievement among recent cohorts are very small (with test scores slightly favoring boys and grades favoring girls) and therefore cannot explain the large gender gap in choice of STEM college majors. Young women have comparable math and science skills as young men, but perhaps have more options to choose from, as they have English and humanities skills that are higher than men's."

worth pursuing, and even with his fairly modest IQ, he was smart enough to find the answer.

To compound the effect of the cultural pressure that discourages teenage girls from studying science and math, the decisions a woman makes in adolescence limit her ability to pursue a career in physics later. Bad enough if a student arrives at college determined to major in math despite not having taken calculus in high school; she might need to spend a year or two catching up, but she won't also need to play catch-up in another field to be able to understand what's going on in calculus. The same is true for chemistry and biology—neither field requires much math. But a freshman won't be able to take advanced physics until she has finished the advanced calculus she missed in high school. The burden of catching up in not one but two subjects, with one of those subjects requiring mastery of the other, is what dooms students of both genders who enter college wanting to major in physics despite having attended less-than-stellar high schools.

Even if a girl finds copies of Danica McKellar's books and feels inspired to excel in math, she can't take a class in advanced algebra if her school doesn't offer one. She might be able to take an AP course online. But she would need to be extremely motivated and well informed to pursue that route. Here's what's so confusing: there's algebra, and then there's *algebra*. When I got to Yale, I knew I had taught myself calculus, so I understood my foundation in that subject might be shaky. But I had no idea the teachers in my regular classes in algebra, geometry, and trigonometry had skipped the more complicated topics in our books. Now, thumbing through the texts used at competitive private and suburban high schools, I realize how much easier my life would have been if I had arrived on campus understanding the behavior of polynomial equations, logarithms, and trigonometric and hyperbolic functions. I didn't know enough to know how much I didn't know. I blamed my deficiencies on my lack of talent.

A friend of mine in the Yale admissions office told me that if an applicant of color wrote on her application that she wanted to study physics even though her transcript revealed a thin background in science and math, the college would accept her and assume the system would do its job and persuade her that she belonged in another field. My friend wishes the admissions office would be honest and say, "Look, we'll let you in, but if you want to major in physics, you'll need to do boot camp the summer before you come, or spend a remedial year catching up." But Yale is too conscious of its status to do any such thing. "We're Yale!" my friend mimics. "We don't have *remedial* anything."

An administrator at another university told me the DUS in math devotes a lot of energy to encouraging students he deems to be talented, but he decides who is talented based on a "flash judgment of the students' status when they come in—did they take college classes while they still were in high school, did they win contests—so of course most of those kids are men or women who come from privileged high schools."

When I ask Cinda-Sue Davis, the director of WISE at Michigan, what she thinks of physics students with weak backgrounds being required to enroll in a math-science boot camp the summer before they arrive, she informs me Michigan already runs such a boot camp, then follows it up with all sorts of extra advising, tutoring, enrichment, and cash incentives during the student's first two years in Ann Arbor. The trouble is that students from weak schools come in needing a course or two in *pre*calculus. Theoretically, they could spend a year catching up at a community college, but the courses wouldn't be rigorous enough to prepare them for U of M. They would need to stay at Michigan a fifth year, and how could the poorest kids afford that? Most freshmen in the program earn higher GPAs than might otherwise be expected. But they fall down the second year, when the material gets harder. "They're often working two or three jobs to be able to stay in school," Davis says, "so they don't have as much time to devote to their research projects as their more fortunate classmates, which means that, given the emphasis on research experience when it comes time to apply for an engineering job or graduate school, in the end we still are losing them."

If female students do make it past their introductory courses, they may continue to face the teasing they assumed they had left behind in high school. As director of WISE, Davis sees few cases of outright harassment. It's more that women complain about their peers behaving boorishly and their professors doing little or nothing to intervene. Engineering is team-based, and the women tell Davis, *The guys on my team tease me. They make me take notes. They don't let me do anything.*

Not long ago, a bunch of female grad students came to see her because their class had been invited to dinner at a faculty member's house and the men in the class had pasted the head of a female physics professor on the cutout of a nude body, and the men couldn't understand why the women got upset.

And then there's the reluctance of most scientists to encourage students to persevere despite their substandard preparation in math, their clumsiness in the lab, their difficulties with Electricity and Magnetism or Thermodynamics. When I tell Meg about my former professors saying they don't advise anyone to go on in physics or math because it's such a hard life, she blows raspberries. "Oh, come on! They're their own bosses. They're well paid. They love what they do. Why not encourage other people to go on in what you love?"

Not that she believes academia is the only career for someone with a physics degree. "I'm torn between that and encouraging people to go on to grad school," she says, then sits scowling, dissatisfied with her answer. "Maybe I need to revise what I've been advising people. I give many, many alumni talks, and there's always a woman who comes up to me and says the same thing you said. *I wanted to become a physicist, but no one encouraged me. If even one person had said, 'You can do this . . .'*" She laughs. "Women need more positive reinforcement, and men need more negative reinforcement! Men wildly overestimate their learning abilities, their *earning* abilities. Women say, 'Oh, I'm not good, I won't earn much, whatever you want to give me is okay.'"

Meg herself needed more positive reinforcement than she got. Her father told her that she was smart, but her need for validation was really high, and nothing anyone ever said to her was enough; she would forget the positive things people said and remember the negative things for years. That's pretty much the way it is with most women she knows. When she was a postdoc, she was friends with a woman who was extraordinarily smart and beautiful, but "there was a terrible mismatch between her abilities and her own estimation of those abilities." Over and over, Meg has seen women with above-average abilities drop out of science, while "more men of mediocre talent are determined to prevail . . . and they do."

It would be one thing if the weed-out system worked, she says. Then, even if the system weren't kind, you could defend it. "But it just isn't true. Lots of mediocre people have shitloads of confidence, and there are many brilliant people who have no confidence." Even if you don't start out being confident, "you can grow the boldness you need to be a good scientist." Look at you, she says. "Given what you accomplished at Yale as an undergraduate physics major and what you've accomplished since then as a writer, I can't think of any reason you couldn't have done what I did and ended up where I am today," a statement that fills me such a mixture of pride and regret tears spring to my eyes. Meg describes a student who told her that she wasn't good enough to go to grad school, even though the student had earned nearly all As at Yale, which has one of the most rigorous physics programs in the country. "A woman like that didn't think she was qualified," Meg muses, "whereas I've written lots of letters for men with B averages."

She can't lie and say getting a PhD is easy. "It *is* a grind. When a young woman says, 'How is this going to be for me?' I have to say that yes, there are easier things to do. But that doesn't mean I need to discourage her from trying. You don't need to be a genius to do what I do. When I told my advisor what I wanted to do, he said, 'Oh, Meg, you have to be a *genius* to be an astrophysicist.' I was the best physics major they had! What he was really saying was that I wasn't a genius, wasn't good enough. What, all those theoreticians out there are all Feynman or Einstein? I don't think so!"

When people ask why so many women go on in medicine and biology but not in physics, math, engineering, or computer science, I tell them you don't need as much math to earn a degree in biology.

One doctoral student said women branch out into "nonpure" fields of physics such as astronomy and oceanography because they like doing things that are less esoteric and have more of an impact on the world, but Meg makes the point it's pretty damn useful to design a hundred-dollar laptop or a cheap but efficient and nonpolluting stove for use in developing countries.

In medicine, women found niches in obstetrics, gynecology, pediatrics, and family medicine before expanding into other fields. They were able to change the culture because male medical students were willing to admit that they, too, would prefer not to destroy their health or ruin their marriages by being on call for days, and they might perform better if they weren't so sleep-deprived they couldn't think clearly, and they might understand more about taking care of other people's children if they spent some time taking care of their own. A physician can, at critical junctures in her career, switch to part-time work without losing her position, while there's virtually no such thing as a part-time tenured professor of physics, mathematics, or engineering.

As to why there are more female chemists than physicists, my hunch is most chemists aren't looking to explain the universe, only to produce a fabric that doesn't wrinkle or absorb odors, a vanilla pudding that tastes more vanilla-y, a bacterium that eats up oil. In academic chemistry departments, where the research is nearly indistinguishable from what goes on in physics, women hold only 13 percent of the faculty positions.

Of course, women aren't the only ones underrepresented in the sciences. One afternoon, I go out for coffee with a young man named Travis, who, like me, completed a bachelor's degree in physics at Yale but didn't go on to graduate school. A sunny, gregarious African American with closely cropped hair and silver-rimmed glasses, he wears a preppy olive shirt and khaki pants. "Even by today's standards, I'm a rarity at Yale, as you were," Travis says, although he seems to be referring less to his skin color than his status as the child of navy parents who "aren't even officers, like most of the other navy kids here." Travis moved eleven times before he found himself attending a huge, poorly funded high school in tobacco country south of Washington, DC. Only 20 percent of his graduating class went on to college, but Travis's parents were the first in their families to earn bachelor's degrees, and they wanted Travis and his sister to do the same.

Seeing that his older sister was the "queen of medicine and biology," Travis turned to physics and math. "Mr. Kelsey, my physics teacher, was the smartest teacher I ever had," Travis says, "and he gave me stuff to read." One day, Mr. Kelsey started waxing philosophical about Olbers' paradox—the question of why, if the universe is infinitely big and filled with an infinite number of stars, the night sky isn't completely light—and Travis couldn't stop thinking about the question. He was the smartest kid at his school, so he took a chance and applied to Yale. When he came to visit, he heard kids talking about studying in study groups and collaborating on research projects. The students seemed a diverse bunch, so he decided to come.

"I did okay freshman year. I wasn't struggling terribly. I mean, I got a B- in Math 120, which should have been a red flag, but I was still adjusting to college life, and I figured if I just worked harder and stopped partying so much I would do better." But second term found Travis failing the electricity and magnetism section of intro physics. He realized the so-called AP classes he had taken in high school "were really just glorified honors classes. I learned to take the tests but not to apply what I knew. . . . My study skills weren't that good. I had to adjust to the idea that tests were such a big part of your grade."

Still, he didn't give up. After repeating introductory physics, he took more intensive courses. He participated in a summer research project designed to support women, minorities, and underprivileged students, but most of the other kids were pre-med, and Travis felt out of place. When he came back to Yale, he began stumbling under the weight of all the advanced science and math classes he was taking, in addition to fulfilling his distribution requirements and trying to maintain a social life. "I felt really out of sorts, tumbling, working really hard, ignoring signs this was a really exclusive club and I wasn't fit to join. I was always catching up, and there never seemed enough hours in a day to get to the point where the other kids seemed to have started."

What kept him from dropping the major was his partnership with a female friend—let's call her Mindy (Travis isn't his real name; both Mindy and Travis are afraid that if they ever return to physics, their detours might be held against them). "With E&M and stat mech I was strewn so far sideways I didn't even know where to start. Mindy would show me where to start, and she would keep asking, 'What does this really mean?'" In the three years they hung out together, Mindy taught Travis to think critically, as did Dave, the tutor with whom Travis and Mindy and two other women worked several nights a week in a study group.

Compared to Mindy, Travis was far more social. "She did nothing but physics. She rarely left her room. She came to Yale with the goal of getting into a really great grad school. . . . She helped me with my problem sets, and I became something of her link to the social world." But Travis wasn't prepared to give up what Mindy gave up for physics. "At the end of the day, it was okay if I didn't understand some concept. Racially, I didn't feel an exclusion from the group of people majoring in physics. But there was a certain confidence of knowing it all. If you don't know the answer, you don't bring it up. Mindy and our group, we *did* bring it up, as opposed to the Caucasian males who did it on their own."

Travis tries to convince me that he isn't as passionate about physics as his friend. But when he describes his senior project in astronomy, his eyes widen and he waves his hands in graceful shapes, creating galaxies in the air. Like me, he struggled to solve the problem his research advisor set him and was unsettled by how often he needed to ask for help. In the end, he came up with a solution. But he's not sure he trusts the results. ("I didn't like the way we needed to fiddle with the math to make things come out right.") Did he really want to work for years on a project that might go nowhere?

What really bollixed up his plan to go on for his PhD was that he got blown away by the physics GREs. "They were just a total mindfuck." Faced with something on the order of 170 minutes to answer 100 questions, he wondered, "Am I not as fast even at *reading* as all these other guys?" He wasn't inept at taking tests—he had scored

780 on the math section of the regular GREs and 750 on the verbal— so he buckled down and studied harder and took the physics GREs again. His score went down. He took the exam a third time and made it into the 35th or 40th percentile. His advisor neither encouraged nor discouraged him (unlike me, Travis thinks advisors should have the guts to tell a student "You can't do this" rather than wait for the student to figure it out), so he guessed he might as well take a shot at applying, but he didn't get into any of the top programs.

Licking his wounds, he took a job in the admissions office. "When I was in high school, literally, the sky was the limit. Majoring in physics at Yale took away my confidence, and I needed to spend some time mourning my loss." Travis brightens. "But I've discovered I'm good at a lot of things. I'm good at talking to people. I'm good at arranging events. I'm good at reading applications and parsing who is good at what." He's the techie on the admissions staff, and he's starting to think his particular set of skills might make him a valuable candidate for a job in venture capital.

Later, I ask Meg about the significance of GRE scores in determining who does or doesn't do well in grad school. Oh, she says, they're not a good indicator of success at all. The professors at a top physics program were asked to rank the students who had finished their PhDs, and in at least half the cases, the rankings showed no correlation with the students' GREs. Even applicants who rank in the 30th or 35th percentile can be successful scientists. For that matter, Meg only scored in the 40th percentile herself.

Nor do extraordinarily high GREs guarantee a student will complete her doctoral degree. "Mindy," the woman who tutored Travis, arrived at Yale confident and prepared. Her mother was a biophysicist who often took Mindy to work and allowed her to amuse herself playing with liquid nitrogen. Having studied calculus as a high school junior, she sought out more advanced math classes at a nearby university. Freshman year, she signed up for the intensive physics sequence, one of only four women in a class with twenty-five men, and she had no trouble keeping up. The part of physics she liked best was quantum, with its "beautiful system of rules," and she felt proud she could excel in a subject other students found difficult. Even though she was doing well on her own, she teamed up with Travis and two female friends to form a group they called "the Minori-team," working through their weekly problem sets with their tutor.

Labs posed more of a challenge. "My lab partner got circuits, so the instructor wanted to move on. I said, 'Wait, you knew that before we started, but I didn't!' But he just kept hurrying on." Still, she felt like a superstar. She had been doing research since freshman year, and some of her results had been included in an article her advisor published. "I think I would have gotten into grad school even if I wasn't a woman," Mindy says. She had nearly all As, great letters of recommendation, and a score in the 68th percentile on her physics GREs. "Maybe it helped a little that I was a woman. But I didn't want to think, 'Oh, crap, I'm not as good as the guys in my class.'"

In choosing a graduate school, she looked for a program where people seemed to be having fun. "That was what I'd liked about Yale. I do feel comfortable with physicists. More comfortable than with other people, really. I took *pride* in being able to hang out with other physicists." Not surprisingly, she got into a top school. But right away, things started going wrong. She broke up with her boyfriend. And the classes were much, much harder than they had been at Yale. "'Oh my gosh,' I thought, 'I worked so hard as an undergraduate, and now the work is even harder. I'm in my twenties, I'm supposed to be having fun, and here I am up until two a.m. doing my math homework. I can't see the physics in the equations, I must be stupid.'"

Given her early success in quantum, she might have preferred to become a theoretical physicist. She was better at math than her "experimental friends" but not as good as her "theory friends." Other people who went on in theory worked harder—at least, Mindy supposes they did—or they were smarter. The experimentalists liked tinkering, liked playing with circuits and computers. "I wish I'd had more opportunity to play with computers in high school. My cousin is doing a computer startup. He bought stuff at RadioShack and built a burglar system. I didn't do any of that as a kid." She got through her classes. "But I wasn't having much fun. And the thought of working this hard for the next four years, then for the rest of my life . . . With problem sets, you knew when you were done. But with research, you *never* knew. You can work for years and years, and the results might still be unpredictable. I mean, you can't pick something you *know* will work and expect people to still care about it."

This lack of certainty led to a nasty fallout with her advisor. "As a graduate student, I needed my advisor to motivate me, and he wouldn't. I wouldn't just take his word for it that the project would work, and he got really angry and yelled at me that I was a graduate student and he shouldn't need to motivate me." The advisor kept yelling at her, and she started to cry. "It was just a reaction I couldn't control." But her crying aggravated her advisor even more.

At the end of her first year, she not only left her advisor's lab (along with the other two women he was supervising), she dropped out of grad school. "I realized I have a nurturing personality," she tells me. As an undergrad, she had enjoyed tutoring Travis; in grad school, she enjoyed taking care of the other first-years. "But there was very little chance for nurturing anyone other than that. You're so isolated. You're completely on your own. You're not motivating anyone, you're not helping anyone." And what was all that isolation for? "If I were successful, ten people would read my paper. If I were very successful, maybe a hundred people would read it. The men seemed to be encouraged by the prospect of publishing a paper. Yahoo! But to me, the trade-off didn't seem worth it. Grad school is miserable for everyone. But it seems to me that women get stopped by their unhappiness, while men seem more able or willing to push through their unhappiness to a future goal."

When we talk, Mindy is only a few weeks into her first term teaching physics at an all-girl boarding school. But she is looking forward to combating the stereotype that physics is very hard, that studying physics can't be fun. "I got really depressed because I couldn't motivate myself to work hard in graduate school. But I can do it teaching, and I'll be able to get the immediate gratification of making a difference in someone's life."

• • •

A woman who dropped out of the graduate physics program at Harvard told me it's difficult to put a finger on why she left. "All these people at Harvard are really trying to create a better environment for women. They're trying, but they're not doing the right things. They need to make there be less emphasis on competition. There needs to be more of an emphasis on cooperation. It shouldn't all be so isolating. And there needs to be more understanding of how women need to be encouraged. It's hard to articulate, especially when you only know your own life. They were doing so much to help me, and it wasn't working, so I figured my not being happy must be my own fault."

Like Mindy, Meg thinks men are more inclined than women to tough out graduate school because society pressures them to find high-powered jobs, while women have an excuse to take the easier route. "Men aren't allowed to doubt their abilities—they're supposed to stay on the path, no matter what. Women are more tempted from the path by people stuff. They value their social lives more than most men. . . . If a woman quits a high-powered career to stay home and raise the kids, she gets a pat on the back. A man quits for the same reason and people say, 'What's wrong? You're wasting all that training? That's nuts!'"

In fact, when I sit down to dinner in Harvard Square with five young women who graduated from Yale with physics degrees, the first topic that comes up is raising children. Three of the women are studying for their PhDs at Harvard—two in physics and one in astronomy—and two are studying oceanography at MIT. Not one expresses concern about surviving graduate school, but all five say they frequently worry about how they will teach and conduct research once they have kids.

"That's where you lose all the female physicists."

"Yeah, it's even hard to get your kid into childcare at MIT."

"Women are just as willing as men to sacrifice other things for work. But we're not willing to do even *more* work than the men—work in the lab and teach, *plus* do all the childcare and housework."

The irony is that Meg and other senior female scientists I talk to don't experience nearly as much difficulty balancing childcare and a high-powered job as these younger women fear. In the past, children did hinder a woman's progress in science. But with changing societal expectations about housework and childcare, the advent of day care centers at universities, and the increased willingness of administrators to accommodate two-career couples, the reality might be that working as a scientist makes caring for a family easier than is true in other professions. What most young women don't realize, Meg says, is that being an academic provides a great deal of flexibility. "And what better way to meet men?" she quips. She met her husband her first day working at the Goddard Space Center. "And we have a completely equal relationship. When he looks after the kids, he doesn't say he's 'helping' me."

When Bonnie Fleming was up for her third-year review, she went to Meg and said she didn't know if she should tell her promotion committee she was pregnant. "Oh, you should tell them," Meg advised. The committee chair had four kids of his own, and when he heard Fleming's news, he said, "That's great! You guys will be great parents!"

Tenured now, Fleming takes her kids everywhere so the younger women will see "you can do this and bring your kids." Besides, plenty of her male colleagues need to leave meetings at five to pick up their own kids from day care, so they are starting to speak up. She thinks a female scientist shouldn't have any problem raising a family if she obeys two rules. First: marry the right guy. Fleming herself married a fellow physicist. If one of them needs to get something done in the lab, if one of them needs to travel, the other takes care of the kids. Second: live by childcare. Raised by a single mom who was an academic, she considers herself proof that sending your children to day care doesn't harm them. Oh, and there is a third rule: "Don't read parenting magazines. They'll only make you crazy!"

As the mother of two boys, Fleming has learned to be extremely effective in using her time. "Having kids has made me a better scientist. How could it not? Having new experiences changes the way you think, the way you exercise your brain." She points to her belly. "And I'm pregnant with a third. A girl."

. . .

If a female PhD doesn't quit to raise children, she faces the long slog of competing for a junior position, writing grants, and conducting enough research to earn tenure and then promotion to full professor. Yet women running the tenure race must leap hurdles higher than those facing their male competitors, often without realizing any such disparity exists. In the mid-1990s, three senior female professors at MIT came to suspect their careers had been hampered by similar patterns of marginalization. They took the matter to the dean, who appointed six senior women and three senior men to investigate their concerns. After studying the data, the committee concluded the marginalization experienced by female scientists at MIT "was often accompanied by differences in salary, space, awards, resources, and response to outside offers between men and women faculty, with women receiving less despite professional accomplishments equal to those of their colleagues." The dean concurred with the committee's findings. And yet, as noted in the committee's report, his fellow administrators "resisted the notion that there was any problem that arose from gender bias in the treatment of the women faculty. Some argued that it was the masculine culture of MIT that was to blame, and little could be done to change that." In other words, women didn't become scientists because science—and scientists—were male.

The committee's most resonant finding was that the discrimination facing female scientists in the final quarter of the twentieth century was qualitatively different from the more obvious forms of sexism addressed by civil rights laws and affirmative action, but no less real. As Nancy Hopkins, one of the professors who initiated the study, put it in an online forum: "I have found that even when women win the Nobel Prize, someone is bound to tell me they did not deserve it, or the discovery was really made by a man, or the important result was made by a man, or the woman really isn't that smart. This is what discrimination looks like in 2011."

Not everyone agrees that discrimination that doesn't look like discrimination actually is discrimination. Judith Kleinfeld, a psychology professor at the University of Alaska, argues that the MIT study isn't convincing because the number of faculty involved is too small, and university officials refuse to release the data. Even if female professors have been shortchanged or shunted aside, their marginalization might be the result of the same sorts of departmental infighting, personality conflicts, and "mistaken impressions" that cause men to feel slighted. "Perceptions of discrimination are evidence of nothing but subjective feelings," Kleinfeld scoffs.

Of course, women who feel shortchanged might all be paranoid. But the most recent studies bear out the factual basis of such perceptions. In February 2012, the American Institute of Physics published a survey of 15,000 male and female physicists across 130 countries. In almost all cultures, the female scientists experienced discrimination in the allotment of funding, lab space, office support, and grants for equipment and travel, even after the researchers had controlled for differences other than gender. "In fact," the researchers concluded, "women physicists could be the majority in some hypothetical future yet still find their careers experience problems that stem from often unconscious bias."

A study carried out by researchers at Yale and published in October 2012 is even more persuasive: it directly documents gender bias in American faculty members in physics, chemistry, and biology at six major research institutions scattered across the United States. Jo Handelsman, the paper's senior author, usually spends her time studying microorganisms in soil and the guts of insects, but since the early 1990s, she has also devoted a significant amount of her time to increasing the participation of women and minorities in science. She long suspected that the same subtle biases documented in the humanities were at work among scientists, but she had no data to support such assertions. "People would say, 'Oh, that might happen in the Midwest or in the South, but not in New England, or not in my department—we just graduated a woman.' They would say, 'That only happens in economics.'" Male scientists told Handelsman: I have women in my lab! My female students are smarter than the men! "They go to their experience, with a sample size of one." She laughs. "Scientists can be so unscientific!"

After moving from the University of Wisconsin to Yale in 2010, Handelsman decided to conduct a study that would either prove or disprove her hypothesis. Excited to be on the same campus as Meg

Urry, whose work on women in science she admired, Handelsman teamed up with Corinne Moss-Racusin, a postdoctoral associate whose interests lie in understanding inequalities of gender, race, and social class. An expert in statistics, Moss-Racusin and her collaborators in psychology, psychiatry, and the School of Management designed a study in which they would send identical CVs to professors of both genders, with a cover story that the applicant recently had obtained a bachelor's degree and was seeking a position as a lab manager. The catch was that half the 127 participants received a CV for a student named John, while the other half received an identical CV for Jennifer. In both cases, the applicant's qualifications were sufficient for the job, with supportive letters of recommendation and the coauthorship of a journal article, but not overwhelmingly persuasive—the applicant's GPA was a respectable but not outstanding 3.2, and he/she had withdrawn from one science class. Each faculty member was asked to rate John or Jennifer on a scale of one to seven in terms of competence, hireability, likability, and the extent to which the professor might be willing to mentor the student, whether by encouraging him/her to persevere as a research scientist or providing extra help if he/she was struggling with a concept. The professors were then asked to choose a salary range they would be willing to pay the candidate.

The results were astonishing:

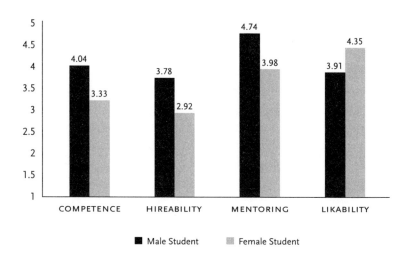

COMPETENCE   HIREABILITY   MENTORING   LIKABILITY

■ Male Student   Female Student

No matter the respondent's age, gender, area of specialization, or level of seniority, John was rated at least half a point higher than Jennifer in all areas except likability (considering that neither candidate was strong enough to be awarded a score higher than 5, the half-point difference seems even more striking). More than that, John was offered an average starting salary of $30,238, versus $26,508 for Jennifer. The significance of the disparity is so great, Handelsman says, when she and Moss-Racusin show the graph to an audience of psychologists, "we hear a collective gasp."

I ask Handelsman if she is surprised female faculty members are as biased as men, but she tells me no, she has seen too many similar results in other studies. Nor is she surprised the bias against women is as strong in biology as in physics or chemistry, despite the presence of more female biologists. As adults, biologists may see women in their labs, but those encounters come far too late to change the biases embedded in their psyches by images they have been absorbing since birth. In a way, Handelsman is grateful the women she studied turned out to be as biased as the men. When she gives a talk and reveals the results, "you can watch the tension in the room drop. I can say: 'We *all* do this. It's not only you. It's not just the bad boys who do this.'"

I ask about objections I have heard that John is simply a stronger name than Jennifer. But Handelsman shakes her head. "It's not just a question of syllables, believe me. There have been studies of which names convey the same qualities to respondents in surveys, and John and Jennifer are widely seen as conveying the same level of respectability and competence." That faculty members report liking Jennifer more than John makes the covert bias all the more insidious. Sensing no outright hostility, a real-life Jennifer might walk away from an interview thinking she had gotten the job, only to wonder what she had done wrong to blow her chances. These results mesh with the findings of similar studies indicating that most people's biases stem from "repeated exposure to pervasive cultural stereotypes that portray women as less competent but simultaneously emphasize their warmth and likeability compared with men"—the very stereotypes being promulgated every night as millions of viewers watch likable, incompetent Penny trade quips with her cold but super-competent neighbor Sheldon.

Of all the biases her study uncovered, Handelsman finds the mentoring result the most devastating. "If you add up all the little interactions a student goes through with a professor—asking questions after class, an advisor recommending which courses to take or suggesting what a student might do for the coming summer, whether he or she should apply for a research program, whether to go on to graduate school, all those mini-interactions that students use to gauge what we think of them so they'll know whether to go on or not . . . You might think they would know for themselves, but they don't. Even the student who is getting 100 percent on every exam doesn't know that everyone else isn't getting 100s, unless the professor tells them." Again, she shakes her head. "Mentoring, advising, discussing . . . all the little kicks that women get, as opposed to all the responses that men get that make them feel more a part of the party . . ."

Handelsman agrees that women need more reinforcement than men, but only because women spend their lives resisting the stereotype of the male scientist. "And the men don't even realize they've been getting all that subliminal encouragement all along, as well as the explicit encouragement. I have male students who say they don't need a role model, they never had a mentor, but they take for granted all the men they've seen as scientists." Nor does Handelsman accept my professors' assertions that they don't encourage students of either gender. "Men do get the encouragement. Everyone talks to their best students about what they'll do next."

Although Handelsman's study tests the response of faculty members to a scientist at the start of his or her career, even when a senior scientist is under consideration, merit may be debatable. "It's not always so clear-cut what a genius really is. There's a lot more room for mediocrity among men, even at the very top, than is true for women. Most of the men on the faculties of even the most prestigious institutions aren't 'geniuses.'"

If you still require proof that a woman can be a superstar and not be recognized as such—even by herself—look no further than Meg Urry. When she was interviewed for the professorship at Yale, the department chair said he would be writing to the top physicists in her field and asking them to name the scientists they considered the very best. If her name came up repeatedly, she would be offered the job. Forget it, she thought. No way in the world would her name show up in those letters. Then the chair called and told her the job was hers. "Let's just say my expectations of the standard they were using were unreasonably high."

Whenever she got a promotion, Urry says, she imagined the guys at the next step must be geniuses. Then she reached their level and saw they weren't any smarter than she was. Even at Yale, she used to listen to the men gossip about other scientists, especially how low their H indexes were—someone's H index measures how often his or her work gets cited by other researchers—and she was sure they must be thinking she wasn't qualified to be their chair. Then she looked up her own H index. "And I realized, 'Hey! I am, in fact, a highly cited author!'" Age helps. "If you stick it out, you discover that your assessment of everyone else is wrong. If you had told me ten years ago that I would be the chair of Yale's Physics Department, I would have told you 'absolutely not.'"

Everyone would agree that discriminating against a gifted scientist like Meg Urry is unacceptable. But some people argue no real harm is done if women choose not to go into science in the first place. David Lubinski and Camilla Persson Benbow, psychologists at Vanderbilt University, spent thirty-five years studying thousands of mathematically precocious twelve-year-olds. Their conclusion? From the start, the girls tended to be "better rounded" and more eager to work with people, plants, and animals than with things. Although more of the boys went on to enter careers in math or science, "the women secured similar proportions of advanced degrees and high-level careers in areas more correspondent with the multidimensionality of their ability-preference pattern (e.g., administration, law, medicine, and the social sciences). By their midthirties, the men and women appeared to be happy with their life choices and viewed themselves as equally successful (and objective measures support these subjective impressions)."

I would be the last person to pressure women to go into fields they don't want to go into. But I don't buy the argument that women know from birth (or age twelve, when they've had plenty of time to be socialized by the culture) that they will be happier in "people" fields. If two groups are presented with two possible paths to follow, one over sharp stones, one over smooth pavement, and one group is wearing shoes and the other, whether by preference or poverty or cultural pressures, walks around barefoot, is it really a choice if most of the people in the group without shoes select the path that isn't laid with sharp stones? I am roughly the same age as the women Lubinski and Benbow followed, and the only reason I tried to be well rounded was that my parents, siblings, and teachers hounded me to smile and put on lipstick, take ice-skating lessons, and (a direct quote from my history teacher, Mr. Burke) "give up that nonsense about studying things and study people."

I do like being around people (if only in the evening). But that isn't the same as wanting to study living things. Do corporate lawyers study living things? Do musicians? Writers spend as much time as theoretical physicists sitting in a room with a pad and pen. The abstractions a cosmologist studies are no less glorious than the notes of a symphony. Most women—and men—decide they don't like math and science long before those subjects reveal their true beauty, a condition worsened by the unimaginative ways in which science and math are taught.

In 2012, the President's Council of Advisors on Science and Technology issued an urgent plea for educational reform to help meet the demand for one million more STEM professionals than the United States is currently on track to produce in the next decade (the shortfall is even more dire if K–12 teachers are taken into account). And yet, even if we strengthen our educational system, we will continue to lose girls at every step as they fall victim to their lack of self-esteem, their misperceptions as to who does or doesn't go on in science, and their inaccurate assessments of their talents.

As I wait for Yevgeniya Zastavker outside her office at the engineering college where she teaches, I pass the time studying a poster on her door proclaiming, "For every boy who is burdened with the constant expectation of knowing everything, there is a girl tired of people not trusting her intelligence," and a report documenting the need for female engineers so we don't end up with airbags that kill women because they were designed to save a man.

Finally, Zastavker comes running down the hall—with her high cheekbones and chic hair, she could be the female lead in an Eastern European art film. I ask about her childhood, and she tells me about growing up in the former Soviet Union, where everyone was required to take physics and she had far more to fear from anti-Semitism than sexism. Only when she immigrated to America as a teenager in 1990 did she encounter the notion that girls shouldn't be interested in science. Desperately poor, still weak in English, she made it to Yale, only to be shocked by the scarcity of women in her classes. As a doctoral student at MIT, she experienced the sort of overt sexism that isn't supposed to exist anymore. (The research assistant assigned to mentor her brought over his chair, "and he put a hand on my knee and said, 'Zhenya, what are you doing here? Your project is trash, this isn't going anywhere, go home and wash dishes.'")

Now a professor at Olin College in Massachusetts, Zastavker conducts research in biophysics and the role of gender in science. Yet even at an engineering school whose faculty and administration are committed to making sure half the students in each class are female, many of the women think they aren't as good as the men. "And strength in numbers doesn't solve the problem," Zastavker says, "because each woman's lack of self-esteem is internal."

Still, the faculty does everything possible to help the women get through. Studies have shown that spatial aptitude is largely a function of experience, for male minority students as well as women (boys consistently outperform girls in tests that measure the spatial skills essential for success in lab work and engineering, but a brief course in how to visualize and manipulate objects virtually eliminates the difference), so everyone is required to take a machining course the first semester. "Everyone is faced straight on with gender differences in the lab. We set them up in co-ed teams and ask them to design a tool or a product. If the gender dynamics get weird, we intervene, and that one intervention early on has a ginormous effect." If she is teaching a regular class and the gender dynamics get weird there, she immediately stops talking about physics and starts talking about gender. She is constantly aware of the "negative micro-messages" her female students are receiving ("Look at the word *engine* at the heart of the word *engineering*, even that puts women off"), so she sends messages of her own—for instance, by tacking photos of her son above her desk.

"Young women worry that if they become engineers, they won't be able to have a family," she says. But the flexibility of teaching at Olin makes raising a child easier than if she worked a regular job. Her husband is a physicist, and he is happy to share the childcare. If all else fails, she can bring their son to the office. She cracks a smile. "Working at a college, I have a whole slew of babysitters to choose from, don't I!"

In Zastavker's seminar about gender and science, I ask the students—all of them female—about their experiences in high school. One woman says she signed up for computer science and was disgusted by the way the teacher joked with the guys but not with her. "They were all just immature, arrogant male nerds trying to out-nerd each other. The teacher kept saying, 'You already know all of this, I'm only teaching this as a review, I know you don't have any questions.' It was just all this joking about sports by guys who wished they could be jocks but were nerds instead."

A young woman named Julie spent two summers working with male engineers. "They made all these awful jokes, and they'd say, 'Julie, you all right with this?' And I'm, like, what was I going to do if I wasn't? I felt competent at the job, but the men were joking about things I'd rather not talk about, and I thought, 'Okay, I can get along with these people, but they won't be my friends.'"

The guys at Olin aren't so bad, one woman says. But male engineers elsewhere are always engaging in pissing contests. "It's always, like, can you burp the periodic table? Thank God we don't have to deal with that on a regular basis here or I would go crazy. It can be fun once in a while to be the girl on the robotics team learning to act the way guys act." Besides, she says, girls at this place are a hot commodity.

"Straight girls, anyway," someone mutters, and when I ask whether it's easier to be a gay engineer than a straight one, the students agree the gender dynamics are complicated. If you are a gay male engineer, that can be looked down on, but if you are a gay female engineer, that's okay. The gay guys stay in the closet, but the women tend to identify themselves early as lesbian or bisexual. "You know, girls in the machine shop . . . that's a pretty butch stereotype anyway," one woman says.

The biggest divide isn't between gay and straight but female and male. If you ask the guys at Olin how they spend their spare time, they'll say, "Building my own computer," while the girls are more likely to sew or cook. "The guys think stuff like that decreases your value as an engineer," someone says. "They would rather run to the dump and bring back some stereo speakers than go running or sit around and chat. Or maybe they'll just sit around all day and play video games in their rooms."

One woman has a dad who is a builder. "You know, Mr. Engineer, Mr. Fix-It—I grew up using tools." But the other seven confess they were apprehensive about the machining course they were required to take.

"I'd never even picked up a hammer before."

"Spending six or seven hours debugging a circuit is definitely not fun."

"A lot more of the guys had experience on machine-shop stuff than the women."

"I felt like the dumbest, clumsiest person on the planet."

"There was this one girl in our group who was terrified of every single machine—sanders, band saws, *everything*. I felt so bad for her . . . she had this terrified look on her face the whole time."

"You know, I heard about there being special toolkits for women—the tools have pink handles, and they were designed for smaller hands, and you keep them in a pink pouch that looks like the purse a prostitute would carry—and I can't figure out if I think that's fantastic or utterly horrible."

"At least if you used that toolkit, none of the guys would steal it!" someone says, and when the laughter dies out, I ask if girls who want to become engineers need more encouragement and support than boys.

Like, duh. "Society tells us all along we're not supposed to be smart, just grow up and be trophy wives and be happy with that, so of course girls need more emotional support than boys."

"Plus, girls come in with less science and math background, then they find themselves in an old white-boys' club. They find themselves in an environment where you're not supposed to show emotion. So of course you're going to be uncomfortable!"

Back at Yale, Meg laughs at my stories of getting thrown across the room by a shock from an ungrounded oscilloscope and not being able to replicate the Millikan oil-drop experiment. Apparently, there is evidence that Millikan fudged his own results. Meg's labs in college were the usual disaster. She hadn't fooled around with batteries or ham radios as a girl. Her male partner did everything while she stood around feeling awkward. Only when she took a more advanced lab—and the guys spent hours trying to get some electronics setup to work, and Meg spent those same hours poring over the circuit diagram, figuring out that her fellow students and the lab assistant had set up the experiment wrong—did she realize she knew as much as they did.

"My boyfriend at the time taught me how to fix things. He told me that the secret to succeeding in lab was 'read the manual.' Most guys *don't* read the manual. They read the first three sentences and stop reading because they think they know what they're doing." She doesn't buy that women are less adept because they didn't grow up messing around with car engines. "I'm soldering things, and I'm thinking, 'Hey, I'm really good at this. I know the principles. It's like an art.' It took me years to realize I'm actually good with my hands. I have all these small motor skills from all the years I spent sewing, cooking, and designing things. We should tell young women, 'That stuff actually prepares you for working in a lab.'"

. . .

I groused for years about my mother not demanding that I get skipped ahead in math and my father telling me—repeatedly—he would rather I earn a teaching certificate in English than a BS in physics. But I never gave them credit for letting me play with Legos instead of dolls, for teaching me to fix the toilet, for showing me how to check the oil on our Impala. My brother taught me how to connect a lightbulb to a dry-cell battery and how to take a pounding on the football field, then get up and go out for another pass. My sister, who earned an MBA from Wharton and went on to become a senior vice president at a major insurance company, taught me to play *Star Trek* on a computer and to refuse to take crap from assholes in a previously all-male field. What scares me is how much less confident I would have been without them.

When my friend Leslie steps out of the Port Authority in Manhattan, I have no trouble recognizing her because she looks exactly the way she did when we were fellow physics majors at Yale, only more stylish and fit. Married, the mother of two sons, she works as a systems engineer at a communications company in New Jersey. We hug, and she tells me again I'm going to be disappointed in her story. Then we spend four hours walking around Midtown, talking nonstop. Not only do we need to catch up on what each of us has been doing since 1980, it becomes clear that I didn't really know much about Leslie even then.

She grew up in New Jersey, where her teachers encouraged her to go on in math. "One of my teachers said, 'Okay, here is a problem that will separate the men from the boys.' I got the answer, and he said I was a man!" Having earned a 5 on the AP exam in advanced calculus and an 800 on the physics achievement test, Leslie applied to Yale. She wanted to be a math major, but she didn't get good advice and took the wrong courses freshman year, starting with an advanced class in linear algebra. "It was totally abstract, we had to do all these proofs, and I hadn't had any proof-based classes in high school. I took analysis and then I dropped it. How could I have gotten such bad advice! I came in strong, I was so confident, but I was really just learning patterns, cookbook stuff, and I took analysis way too soon. What do you know at eighteen? It's all trial and error. My parents didn't know what to advise me. They were struggling themselves."

I ask how she came to study physics—and why she didn't go on in the field.

Well, she says, she took a computer course sophomore year and liked it. But the switch to physics didn't come out of nowhere—she had written in her high school yearbook that she wanted to grow up to be a "quantum mechanic." At Yale, her dream was to go on for her PhD in high-energy physics at Berkeley. But the two summers she worked at the Stanford Linear Accelerator made her reconsider. "I saw all those guys working so hard and sacrificing so much just to understand everyone else's papers, what others had already done, let alone to go beyond it and come up with their own breakthrough. The guys worked all the time, they were so disheveled, just not well-groomed, and I couldn't imagine being in their company." She muses about the miniscule number of physicists who have managed to combine a love for ballet or hiking or history with a career in science. "I would be in physics still if I was so brilliant that I could lead a balanced life."

After working at the cyclotron at Harvard Medical School, Leslie earned her master's in energy policy from MIT, took a job at the Department of Transportation, then a job at the Federal Aviation Administration, and then a terrific job at Bell Labs, which she calls a "magical place." Everyone was friendly and bright and incredibly creative; she even met her husband there. After Bell Labs closed, she moved to a communications company that encrypts data for the Department of Defense. I ask if she minds working for the military—I've seen a YouTube video in which she pleads with President Bush to stop the genocide in Darfur, and she is chair of the Social Action/Social Justice Committee at her temple.

"There's not much else you can do. All the nonmilitary jobs in New Jersey were transferred to China. It's only government jobs that can't be transferred because you need to be a citizen for those."[2] She

---

2. Not long after, she lost the job; later, she enrolled in Hacker School in Manhattan, "coding all day, every day, for three months to become a better programmer . . . working in a community of other software developers who are motivated to do the same."

basically enjoys what she does. But with all the hours she puts in working and commuting and raising her sons, she doesn't have time for tennis. And it bothers her that she gets criticized at work for not being nice or polite.

"Not nice?" I say. "That's nuts! You're one of the politest people I know."

She shrugs. If a woman speaks up at a meeting, she's considered not nice, even though a man can say the same thing and everyone is just fine with it. Like one time, Leslie was asked to review a younger colleague's presentation, and she offered him some suggestions, which he totally blew off. Two weeks later, a male colleague made the same suggestions, and the younger guy told him, "Oh, absolutely!"

The sun goes down, and I walk Leslie back to the Port Authority so she can get home to her family before dinner. Her older son is about to enter his first year as a physics major at Rutgers. He got into MIT, and he loved his visit to the campus—the students were playing Quidditch—but Leslie was afraid all the competition and stress might cause him to stop playing his saxophone and destroy his love of physics. She hopes four years at Rutgers will allow him to gain some confidence and explore interests other than physics before he subjects himself to the pressures of graduate school.

Still, she can't help but envy him. He gets to major in the subject they both love, but with her good advice and the stability of the family she and her husband have provided. "I wish I could do it again," she says. "With what I know now. With me as my own parent."

And my former suite mate Erika? From a clip on YouTube, I learn she is now vice president in charge of market development for a biomedical company in California. As polished and poised as she was at eighteen, she is several orders of magnitude more so now. I am dying to know where she got such confidence. We set up a time to talk, and I ask the question straight out, to which Erika provides the answer I should have learned to expect—namely, that she attended an all-girl high school.

And physics? How did she come to major in that?

"You know how people make choices at that age. It's rarely a thoughtful process. It just seemed to me that if you were really smart, you did physics or math. Getting into Yale was a confirmation that you were smart and you should just keep doing smart things." It never occurred to her that women shouldn't do physics. "Gender never came into it until I got to Yale. And even at Yale, I never really thought that much about gender, except that there were so few women in our classes."

Like me, Erika hadn't taken real physics or calculus in high school, and she got a terrible grade on our first physics midterm. Unlike me, she didn't let the low grade faze her. She was on the Silliman social committee. She always had a boyfriend. "There were not that many times I buckled down, so I never would have thought I was being unjustifiably looked down on. It was a fluky reason that I went into physics. It wasn't a passion for me the way it was for you."

Erika was fine with real-world stuff like balls rolling down inclines. But she hit the wall when she started taking courses in which the material was more abstract. Sensing she would be more comfortable taking classes in engineering, she signed up for acoustics and computer science. "I remember one of the problems we solved was how to launch and land a rocket on the moon. You had to take into account the trajectory of the rocket, its speed and how much fuel, and you couldn't miss the moon or crash into the surface. It's an easy problem to solve now, but back then—you remember this, Eileen—you had all these stacks of punch cards." That's when she realized she was "more curious at the engineering and systems level—how do you balance different parts of a complex system, how do all the pieces work, if you push it here, what will pop out there."

Her only regret is no one counseled her to switch majors. "It would have been incredibly helpful if someone had said, 'Young lady, why don't you go into engineering! Your mind doesn't go *this* way, maybe it goes *that* way.'" But she didn't seek anyone's advice, so she didn't think to pursue a master's in engineering—or, later, her MBA—until she was several years out of Yale. "In my twenties and thirties, I thought people were only doing business because they couldn't do science. Once that bubble burst for me, I realized that if I toted up the smartest minds I knew, they were rarely scientists. The thinkers I found who were smarter than me were all in the business world." After working as a consultant for several start-ups and venture capital firms, Erika joined the biomedical company where she now works. "It's a real hoot to have all these different strings—chemistry, physics, biology, business—to have all these different strings get pulled together." She has encountered "a few gender things" in the business world. But they didn't deter her because "I'm confident as a person."

I can't tell whether I admire Erika or hate her. Maybe that's my excuse for asking if she is married, if she has children.

No, she says. No kids. But she and her husband, who studied physics and math at Grinnell and earned a law degree from Harvard and now specializes in intellectual property rights, have been married for twenty-five years and share a house in Colorado. If all of it—the confidence, the success, the personal happiness, the lack of regret that she didn't pursue a career in physics—if all of that is an act, the act has me fooled.

"The goal is not to have more women physicists," Erika says. "It's to find the people for whom physics would be a good fit and open them to the possibility. There's just as much nine-to-five drudgery in physics as in any other field, and there's just as much opportunity for creativity outside physics."

I know the interview is supposed to be about Erika, but I can't resist following up on her comment that she perceived me as being more dedicated to science than she was. What else does she remember about me from those days?

Erika hesitates, and I tell her I can handle the truth. "I guess I would say I thought of you as insular. Looking in more than looking out. Dedicated and serious as a student in a way I wasn't. I thought you had a mission, a duty to do well in physics. You had made a choice, and you were at Yale because you really wanted to learn. To learn physics. As I remember, you didn't do a lot of things other than study. You didn't talk much. You were a little unsure of yourself, nervous about your interactions with other people."

When I tell this to my son, he says, "Mom, that can't be possible. You're so outgoing! And you have so many friends. You run this really big writing program, and you give readings all the time. Aren't you upset by what she said?"

"Why should I be upset?" I say. "Everything she said was true."

. . .

It's a Friday afternoon and I am relaxing at a picnic for the Yale Physics and Astronomy Departments, surveying coolers of beer and tables crowded with barbecue fixings, pies, cakes, and a plate of brownies Meg's husband baked that morning when he realized she had overslept and wouldn't have time to bake them. Behind us rise two mammary-shaped domes, one of which houses the eight-inch refractory telescope Yale purchased in 1882 to track the transit of Venus across the sun and now uses for student research, the other a planetarium to introduce local kids to the wonders of the stars. The weather is glorious, and students and professors who look nothing like the geeks on *The Big Bang Theory* (and yes, a few who do) are enjoying their annual fall get-together before the weather turns cold and the semester heats up and they have so many lectures to plan and papers to grade and journal articles to referee that they don't have time to sip cabernet from a plastic cup and chat with a colleague who has just returned from a semester's leave at an observatory in South America. Children roll down the hill, young men spread mustard on their hotdogs while balancing toddlers on their shoulders, and I sit at a table with four young women Meg wanted me to meet.

"Face it, grad school is a hazing for anyone, male or female," one of the postdocs says. "But if there are enough women in your class, you can help each other get through." In her case, she was lucky to be one of three women who shared an office. "We spent hours and hours every day sitting in that office, knitting, programming, thinking, talking girl talk. We got lots done, and we kept each other sane."

Another postdoc says she stayed sane by boxing. "You can't punch your computer. You can't punch your advisor." She throws a punch in the air and grins.

I look at these four women—one black, two white, one Asian by way of Australia—and I can't help but ask how they made it so far when so many other women gave up their dreams.

"Oh, that's easy," they chorus. "We're the women who don't give a crap."

Don't give a crap about . . . ?

"What people expect us to do."

"Or *not* do."

"About dressing up and wearing makeup and getting our hair done."

"Or"—this from the black graduate student, who wears dangly earrings and a stretchy low-cut shirt—"about men not taking you seriously because you dress like a girl. If you're not going to take my science seriously because of how I look, that's your problem."

One time, one of the postdocs and her female lab-partner wrote up a lengthy report about some experiment. When two male classmates asked if they could use the women's write-up as a model, the women said sure. The men weren't cheating—they had carried out their own version of the experiment and would be filling in their own data—but the two teams turned in virtually identical reports. And yet the men got an A and the women got a C. Willing to give their professor the benefit of the doubt, they waited until the second time this happened before confronting him. "We almost felt sorry for him," the postdoc says. "He tried to justify the grades. 'Well, um, their reports are more . . . more . . .' It was totally pathetic. We told him straight out the reports were identical and he had given us a C because we were women, and you could tell he just didn't want there to be a conflict. He changed the grade, just to make us happy. But what did we care, we knew the truth."

The African American woman did her undergraduate work at a historically black college, then entered a master's program at Fisk designed to help students of color develop (according to the program's website) "the strong academic foundation, research skills, and one-on-one mentoring relationships that will foster a successful transition to the PhD." Her first year at Yale was rough, but her mentors at Fisk helped her through. "As my mother always taught me," she says, smiling, "success is the best revenge."

Each of the four women has a partner or a husband. "Are you kidding?" one of the postdocs says. "Majoring in physics is the best way to meet men. Smart men." Her toddler pulls away from his father and propels himself onto her lap. She pops out a breast and nurses him. "Who wants to be with hot guys whose eyes glaze over when you try to explain the Doppler shift? Not me."

CHAPTER FIFTEEN

Parallel Universes

In *Has Feminism Changed Science?*, an astrophysicist named Andrea Dupree makes the case that what keeps women out of cutting-edge theory is "that extra bit of chutzpah or aggressiveness or assertiveness." To be a theoretical physicist, she says, requires "a certain sense of ego and the ability to be verbal, to be articulate, and to be aggressive. . . . Theorists love to rank all the other theorists in the world."

Would I have had enough chutzpah to make it as a theoretical physicist? I will never know. But I grew up in a place where chutzpah was ladled out to every child, female and male, along with the chicken soup and borscht. In my late forties, I learned to play such aggressive tennis that any opponent careless enough to send a poufy shot my way cringed in fear that my overhead or volley might ricochet off her head. In my fifties, I found a pro who helped me to appreciate the beauty of hitting a perfect stroke and the joy of playing the game as it is meant to be played, rather than caring so much whether I win or lose. Which, of course, greatly improved my ability to win a game.

Also, you could substitute *writer* for *theoretical physicist* in Andrea Dupree's quotation and have a perfectly accurate statement.

. . .

When I was young, I was certain I was the only woman who thought and felt the things I thought and felt. Now, after so many years of loving and being loved by extraordinary men and women who know me for who I am, after decades of reading and writing fiction, after realizing that I can feel as comfortable in a slinky, revealing cocktail dress as in a sweatshirt and jeans, after giving birth to a son and watching him grow to be a man any writer or physicist would want to love, I know I have thought and felt very little in my life that most other women haven't also thought and felt.

If a talent for scientific thinking can be inherited, my son received a double dose of the gene. He taught himself to multiply while sitting in the tub staring at the ceramic tiles, then spooked his preschool teachers by asking them to give him "times problems." Not long after that, he said, "Mom? Did you know that if you add 'finity and 'finity, you still get 'finity?" Then he laughed at how ridiculous 'finity seemed to be.

But he also has parents who encouraged him to think. One afternoon, the light from the kitchen window shone in such a way that it isolated the drops in a thin stream trickling from the faucet. "You can see the bubbles in the water," Noah said. "Is water made of bubbles?" Prompted by his questions, I explained how hard it is to predict the exact instant a succession of drops might switch to a steady flow, how the equations that govern the behavior of water are so complicated scientists still haven't solved them. He stared a while longer. "It's like those photographs. They use a strobo . . . strobo-something." A stroboscopic light? He flapped his hands, as he did whenever some new thought excited him. "They used a stroboscope light to take pictures of this drop splashing in a puddle. That's what's happening now, isn't it, Mom?"

I could never have conceived such a thought at his age. Then again, my own parents would have snapped the faucet shut and told me to stop wasting water.

I admit, I wanted my son to become a scientist. Science is a more communal art than writing. The frustration is the same, but scientists waste far less time questioning that their work will benefit humankind. Most scientists are optimists: surely the same human minds that mastered calculus and quarks will come up with solutions to poverty, war, and global warming. Writers meditate obsessively about conflict and despair. "He might become a great writer if some woman breaks his heart," I once commented about a student. No one would say such a thing about a physicist.

And yet, would I really want my son to live in ignorance of the romantic desire that can disrupt the neatly planned boundaries of a career? His father and I could have sent Noah to a private high school. But he might never have come into contact with the sorts of kids I met growing up—poor kids, mean kids, kids who couldn't grasp the concept of a cosine but played jazz or sang harmonics with a tender, inspired insanity that broke your heart. In the end, Noah's father and I confined ourselves to helping our son acquire the ability to sit quietly and pay attention, the talent for wonder, the confidence to ask a question and pursue its answer. We raised him to be happy with who he is.

As it turns out, what he is passionate about is making sure everyone grows up with the opportunity to develop his or her talents and pursue whatever passion he or she wants to pursue. "But he's so smart!" people say, as if making sure everyone has the opportunity to develop his or her talents is a waste of my son's intelligence.

According to *Why So Few?*, only about 4 percent of the American workforce is employed in science, engineering, and technology. Yet this relatively small group is critical to innovation and productivity. Given that most jobs in science and engineering are highly paid, it makes sense to cultivate a society in which people of both genders can gain the skills they need to succeed in such careers.

But we also need people who can help us figure out what to think and feel about all those remarkable innovations. I have spent the past twenty years trying to win respect for the humanities at a university where the lion's share of resources goes to the medical school, the business school, the engineering school, and the football team. I have spent much of my teaching career trying to convince science and engineering majors that the humanities provide us with a way of thinking as well as feeling. I have spent most of my adult life being condescended to by the men I meet at parties . . . until I slip in that I have a physics degree from Yale.

After I finished this book, I sent the manuscript to Larry Summers. Here's what he e-mailed back, in a message he described as "heartfelt":

> Eileen—thanks for sending me your manuscript. My remarks on women and science generated much heat—if they helped stimulate your introspections and reflections they shed light as well. I certainly understand many aspects of the issue better for reading what you have written. We all want great opportunities for all, and certainly as you demonstrate the world has a long way to go.

Late one night in the spring of 2011, a Yale physics and astrophysics major named Michele Dufault was working alone in the machine shop in the basement of Sterling Chem when her hair got caught in a rotating lathe and pulled her into the machine, crushing her neck. Meg Urry was devastated. She had known Michele since she was a wide-eyed freshman in her astrophysics seminar (Michele, the only freshman, earned an A). And in her four years at Yale, Michele had devoted herself to mentoring younger physics students. At Michele's memorial service, Meg recounted the meeting she had held to discuss the next year's Conference for Undergraduate Women in Physics, the real purpose being for Michele to pass the baton to the younger women. "They weren't sure they could handle everything, but Michele just encouraged them and offered to help. She would be available whenever they needed her, she said, including next fall."

Meg sent me a transcript of the remarks Michele's classmates offered at her service, and I was struck by the comments from a young man named Zach, who said Michele had regularly stayed up past three in the morning to help him and his friends finish their problem sets, long after she completed her own. Another friend, Joe, described the time he and Michele spent hours in a Houston aircraft hangar on a "smoldering summer day" arguing about laser containment. The younger members of Yale's Microgravity Research Group were off preparing for their adventures in NASA's C-9 "Vomit Comet," leaving Joe and Michele "struggling to assemble our recalcitrant experiment. Michele would never compromise her deeply held beliefs, not on moral matters and not on scientific and technical issues. So, our debate on that tumultuous afternoon was intense." But when the rookies on the team returned from a battery of training exercises looking bedraggled and drained, Michele insisted they trek to Ben & Jerry's for ice cream. Without Michele, Joe said, "not only would our hardware have been in shambles, but also the team's morale would have cratered."

Other than a tendency to walk and talk in our sleep, I'm not sure how much Michele Dufault and I had in common. I can't imagine helping my male classmates with their problem sets, or arguing about the best way to set up a laser. Even at Yale, Michele found time to play the saxophone, raise money for homeless shelters, go rock climbing, and play Frisbee in the rain. She had long, lustrous brown hair and a wide, fluorescent smile—in the movie version of her life, she would be played by Danica McKellar.

And yet, Meg is sure if Michele and I had met, we would have discovered we were on the same wavelength. In fact, Meg is convinced Michele was the young woman who stayed after the master's tea and told me how much she hated when her sister introduced her as an astrophysics major, because the boys would turn away.

"I feel like the resonances you talked about—somehow your frequency and mine and hers must be harmonic," Meg wrote me in an e-mail. "Michele once asked her mother why she didn't have a boyfriend. She was beautiful and clever and the loveliest, most thoughtful person you could imagine. The men at Yale must be morons."

To wish I had been born in a different era—say, in the late 1980s, when Michele Dufault was born—strikes me as useless. I might equally thank the stars I wasn't born in medieval times.

But what if my life had taken a slightly different path? What if I had grown up to be Meg Urry? Would that have made me happy?

The person I am now would tell you no. But the person I was as an undergraduate—and the person I would have continued to be had I stayed in physics—would admit that Meg Urry is exactly who and what I wanted to grow up to be.

If an infinite number of universes exist and everything that might happen in a person's life occurs in at least one of those universes, in one of the universes running parallel to the one in which this Eileen Pollack lives, there lives an Eileen Pollack who did become a theoretical physicist. In which case, I hope that Eileen Pollack loves her life as much as this Eileen Pollack loves hers.

To wish away the circumstances that prevented you from becoming what you most wanted to be is to wish away the self you are now. Maybe if I had received the support I needed to become a physicist, I wouldn't now feel special for having achieved all that I have achieved despite the obstacles in my way.

Then again, maybe I would feel more special. Maybe I would take pride in my actual achievements instead of my endurance in the face of so many obstacles.

All I know is I wish I hadn't wasted so much time and energy trying to win the approval of men who didn't care, or even notice, what I was doing. And I wish the men who would have taken pleasure in my success—Barry Talkington, John Hersey, and my father—might have lived to see me achieve the happiness and success they hoped I might achieve.

Which only goes to prove that if you want to become a physicist—or anything else—you need to do it for yourself. You need to do it for the little girl who couldn't stop thinking about how everything that exists evolved from nothing, how the first human beings learned to speak inside their heads, whether time would exist if no observers existed to record it, how a ray of light sniffs out the fastest path to follow, how an electromagnetic wave might appear if it were traveling in two or four rather than three dimensions, how an infinite number of infinitesimally tiny slivers beneath a curve can be integrated to find its area, or how an infinite number of infinitesimally tiny fractions of a human life can be combined to create a whole.

EPILOGUE

The Sky Is Blue

On October 6, 2013, an excerpt from this book ran in the *New York Times Magazine*. The number of women who wrote to say they connected with my story—and wanted to share stories of their own—was both inspiring and disheartening. At the project's start, I had intended to use my experience to help explain the scarcity of tenured female physics professors at elite institutions such as Harvard. In the end, I discovered I was speaking for women who had struggled to pursue careers in science, math, engineering, and computer science at every level, as well as women attempting to succeed in medicine, architecture, finance, and law.

I wasn't surprised so many women my age wrote to recount the humiliation they had suffered. Those stories, though they moved me to tears, confirmed what I already knew. And I was thrilled so many men wrote to express their awe for what their mothers and grandmothers must have endured in earning their degrees in science, their sympathy for the struggles of their female students, their outrage on behalf of their wives, and their anxieties for their daughters' futures.

What broke my heart was how many young women echoed the complaints voiced by those students whose baffled outrage erupted at that tea at Yale. A high school senior at a competitive private academy in Connecticut said she is one of only three girls in her AP Physics

234

class and one of only four in Linear Algebra. She and the other girls always feel like the dumb ones. They are afraid to ask questions for fear of being judged by the boys, who appear so much more confident. A young woman who called to interview me for her school newspaper said that in preparing for a robotics competition, her female friends had been assigned to the decorating committee, which caused one of those friends to quit. Another woman complained that a professor taught her engineering class the mnemonic device "Bad Boys Rape Our Young Girls But Violet Gives Willingly" to help the students remember electronic color coding. The mother of a senior at an Ivy League university said her daughter had suffered such severe exclusion and discrimination from her fellow (all-male) math majors, she almost didn't survive her freshman year. Even though the daughter's classmates were all assigned advisors in the Math Department, the young woman was assigned a law student as her advisor.

Other correspondents wrote to say that, like me, they received so many messages they didn't belong in science that they simply drifted away from the career they had intended to pursue. A former physics major remembered a textbook on optics that had photographs of a woman's chest from a "breast movement study" to demonstrate some technical point. When she showed the textbook to the only female professor on the faculty, the professor grew flustered and did nothing about the student's complaint. The student ended up pursuing a career in business. "So many family members and friends who encouraged me could not understand why I was giving up something that was so rare to be able to do and I felt like I was somehow letting all women down," she wrote. "Your article brings me to tears with the memories of what at the time I just tried to ignore—but a more accurate word is what I 'endured' in getting that degree."

Another woman made it into a doctoral program in applied math before dropping out. Even though she had graduated summa cum laude, Phi Beta Kappa, with a bachelor's degree in math, she faced much of the subtle hazing and lack of encouragement I described in my article. (One of her professors refused to learn her name, referring to her instead by the name of the only other woman in the class.) As for those female scientists with the stamina to remain in their fields and attain the highest rank, many described themselves as worn down by the continual need to fight for respect, not only from male professors and colleagues but also female colleagues, students, and staff.

A number of female engineers wrote to describe the difficulties they face in academic and corporate settings. When one woman visits engineering classes at the local high school and college, she rarely sees girls in the advanced classes, and when she does, their male teachers often treat them with scorn and assign them to be record keepers rather than group leaders. Another engineer recounted a meeting with her freshman advisor at UC Berkeley, who tried to persuade her to switch to biology. Engineering school is brutal for everyone, she said. Two of her male classmates committed suicide. But it's even harder for women because they get less respect.

In one sadly funny message, a female inventor noted she had been patenting life-saving medical devices for a quarter of a century, yet constantly found her progress hindered. Her first patent attorney informed her that her gender devalued the technology she had produced. Later, the head of manufacturing at a company where her firm had pitched a product stopped negotiations because he learned the inventor was a woman.

Even male engineers felt compelled to corroborate their female colleagues' complaints. An MIT graduate with twenty-five years' experience in industry wrote to say that more than 90 percent of his colleagues are male, but the women are as good as the top 20 percent of the men. Male engineers seem able to tolerate a marginal male colleague but will bash a female engineer who demonstrates the same level of skill. His conclusion? The only women who last in his field are the absolute best.

Even in such female-friendly fields as medicine and biology, women struggle to achieve equality. Female cardiologists and surgeons remain rare, and no more than 16 percent of the physician executives in our health-care system are female. A woman attending a symposium at Harvard Medical School celebrating the Warren Alpert Foundation Prize wrote that of the forty-five biologists and doctors awarded the prize since 1987, only one was a woman (the next year, another three men won the prize), and of the eight speakers at the symposium, all were male.

Not surprisingly, women in economics, architecture, and invest-ment banking wrote to say they encounter the same obstacles as fe-male scientists. An economist remembers querying her instructors in graduate school and receiving flippant responses. She went to see a professor with a male classmate who had similar uncertain-ties about the material. The woman asked a question, at which the professor turned to her male classmate, noted that the question was an interesting one, and proceeded to explain the answer to him. An-other correspondent said she had spent the past seven years in the male-dominated fields of investment banking and private equity, and even though she was as talented as the vast majority of her col-leagues, she wasn't naturally aggressive and found it exhausting to continually put on that facade at work. Similarly, a woman in cor-porate finance wrote that she is a funny, laid-back person, but if she is to be taken seriously at meetings, she needs to be belligerent. She wishes she felt as confident as her male peers. When a mathematical question comes up, she is able to solve the problem in her head—not surprising, given she has an Ivy League MBA—but she doesn't trust her math skills, or she worries she doesn't really know what she's talking about. Then someone gives the same answer, and she gets an-noyed at herself for keeping quiet.

This is the theme that resonated most with female readers: women don't know how good they are at science or math because no one tells them. One woman scored highly on the Math Association of America exam. But none of her teachers acknowledged her talent. Girls aren't weaker at science or math, she said. It's just that we place more value on a pat on the head than we do on our own responses. She wishes her father, an aerospace engineer, had talked to her earlier about how much he loved his career. If he had, she might have pur-sued engineering.

One of the most poignant examples of the ways in which women misread the silence that surrounds their achievements came from a former math and physics student whose brilliance seems undeni-able. As a teenager, she represented the United States in several ma-jor mathematics competitions and won international recognition for a pair of proofs. This was just how her brain worked, she told me; math was what she did for fun. Encouraged by her teachers, she went on to study physics at a top research university, where she earned all As. One of her teaching assistants was so impressed he got her a position in his fluid dynamics lab. And yet, she thought she was flailing. Why? The lab director brought her to a conference, then refused to allow her to present her work, even as he encouraged all her male collaborators to present theirs. When asked why, he said a student needed to be "really bright" to present a paper. Later, the student visited her linear algebra professor, who told her explicitly she didn't have what it took to go on in math. She knew her grades were good, but she assumed the other students were doing better. In retrospect, she realizes neither her lab director nor her math professor had any idea who she was, what grades she was earning, or what she was capable of achieving. In judging her chances for success, she relied on those best placed to tell her. They suggested she was average. She did not go on in physics.

Several male professors wrote to say they go out of their way to offer support to their brightest female students, even as they puzzle as to why these students lack confidence. More commonly, male professors wrote to express their impatience, even anger, at women who exhibit "self-esteem issues," who blame their surroundings for their failures, who think they are entitled to more encouragement than anyone else. Women scientists are so difficult to work with, one man raged, not even women scientists want to work with women scientists!

To which I would offer the testimony of a woman who, despite her good grades in science and math and an internship in materials science that resulted in a publication, gave up on her intention to become a chemist. No one harassed her to leave the field, she said. But neither did she feel supported. She didn't understand her options, and in the absence of mentorship or encouragement, she dropped out. "And before you go yelling about how women shouldn't get 'special' treatment or support," she wrote, consider that she was a young girl who had never in her life had anyone suggest that she might be something other than a teacher. No woman in her family had any kind of career. She now works as a financial analyst, but she still reads scientific articles with interest, "and something in me hurts a little when I do."

One of the funniest comments came from a male reader who thought it strange I so readily concluded I wasn't smart enough to pursue a career in science. What was I, he demanded, a hothouse tomato? If I chose to leave science, I should take responsibility for my actions, stop blaming others, and move on. If he had waited for his superiors to encourage him to succeed, he said, he would still be standing outside his professors' and managers' doors.

This reader's scorn for my lack of courage made me flinch. But I had to laugh at the epithet he chose, with its old-fashioned, sexual connotations (my father used to bestow the not-very-complimentary "she's some hot tomato" on any female endowed with a curvaceous shape, a flirtatious personality, and a tendency to show off both) and its reminder of the tomatoes I grew in my parents' attic with hydroponics. The criticism made me realize how much encouragement I did receive. Maybe Professor Parker didn't urge me to apply to graduate school, but he nominated me to attend that conference for physics majors interested in careers in theory. And even though Roger Howe waited thirty years to tell me that my senior thesis was exceptional, would he have agreed to be my advisor if he hadn't respected my work in the two classes I had already taken? Would he have asked me for a copy of my thesis so he could publish it? I want to shake my younger self and say: *Really, Eileen? None of that was enough?*

But then, it's easy to chide someone for being a thin-skinned tomato while overlooking the loamy compost of encouragement with which society nurtures young men by providing them with so many images of male scientists and pushing them to tough out the difficult times, take risks, get the best education possible, and earn a good living to support a family, just as it's easy to discount the withering effects of all the teasing and belittlement that undermine a young woman's confidence. Maybe men would be as insatiable for praise as women if their parents and teachers starved them for compliments about anything except their looks, their deportment, their willingness to neglect their needs in favor of fulfilling the needs of others. Maybe women shouldn't need to hear "you're good" so many times, the way we shouldn't need to ask our partners, "Do you love me?" But who doesn't need to hear "I love you" at least a few times at the beginning of an affair, let alone a marriage?

No one is asking professors to encourage students who aren't talented. If a young person lacks a skill, what she needs is an honest assessment of that deficiency, coupled with detailed advice about to how to remedy it. Which, I am pleased to report, is exactly what a number of male science and math professors do. One woman wrote to say the advisor for her senior project at Yale spent even more time advising her on where to apply to graduate school and connecting her to potential mentors than talking about her research. She overheard a girl ask if this professor thought she was prepared to take his class. He responded that when a man asks if he is prepared to take a class, the answer is almost always no. But when a female asks if she is prepared, the answer is almost always yes. At first, the professor's response struck this woman as weirdly biased. But then, every time she caught herself wondering if she was ready or good enough to move ahead, she realized this meant she *was* ready. Now, she is pained to hear highly competent young women express their insecurities, if only because it's hard to preserve her good opinion of someone who constantly reminds everyone that she doesn't have a good opinion of herself.

I was especially grateful to readers who bolstered my assertion that women's feelings about their appearance or their ability to attract a man contribute to the lack of female scientists. My editors at the *New York Times*—both female and male—kept urging me to cut my references to dating and clothes and qualify any such comments by adding the phrase "as superficial as this might sound." Yet I couldn't discount my experience that I needed to minimize my femininity to be regarded seriously as a physicist. If this was true of someone who grew up in an era when women spurned makeup and revealing clothes, how could it not be true for women today, who are pressured to dress like porn stars? Why should I condemn as superficial a woman's concern for how she looks or whether she will find a man to marry her if she receives messages from every side that her appearance and romantic status constitute the basis of how she will be judged?

In the end, I won the argument by asking my editor how he would feel if he were required to wear a dress to the newsroom every day. And yet, I braced for a deluge of complaints from readers who chastised me for portraying women as superficial. Instead, one of the first e-mails the *Times* received came from a woman who challenged any man to work in an environment where the main topic of conversation was child rearing, where all important after-work socializing took place in a nail salon, and where they knew their attractiveness was being rated every time they left the room. She bet they wouldn't last a week.

Another reader described the day her sister-in-law showed up for an advanced math seminar, only to be told she must be in the wrong room, an assumption the professor announced in front of a class full of young men. He was shocked when she corrected him—not at his own sexism but at the possibility a beautiful young woman should be taking an advanced math class. A female professor at a prestigious Southern university wrote to say her male students regularly ask her out on dates. Even though she has a reputation as one of the toughest professors on campus, they question her authority, tell her how to teach, and try to trip her up in lectures. (They never succeed, she assured me.) In every class, whether graduate or undergraduate, a group of male students giggle and act disruptively. Would she recommend an academic career in a mathematical field to a young, attractive woman? Probably not, she said.

One perceptive teacher wrote to say he sympathizes with his female students who pick up the message that girls, by virtue of their hobbies or the way they dress, have no place in a physics or mathematics classroom. As a physics student, even he found himself baffled by all the references to football and basketball and was constantly asking his male friends questions such as, "What is hang time?" Recently, he showed a group of colleagues a question from a seventh-grade math book that involved calculating baseball averages. No one saw anything amiss. Then he showed them a question that involved analyzing data to determine what size pantyhose a woman should buy. Both genders expressed dismay because the pantyhose question felt "weird."

Other readers attributed women's discomfort with science to the rigid social codes that get imposed on them in junior high . . . and enforced by other girls. Science is not feminine, one reader wrote, and if you don't toe certain gender lines in grade school and middle school, you get ostracized. As much as this reader's experience confirms what I suffered in junior high, I agree with the reader who pointed out that girls enforce these norms because such rules are ground into them from the moment they are born.

Of course men exist who are willing, even eager, to date women who are smart. One male scientist sent me a postcard to the effect that finding a woman with whom one can share a deep common interest in a STEM field "is heaven forever." But my sense is most American men are still intimidated by any female they perceive as smarter. Or they prefer women who are smart in less traditionally masculine fields. In publishing this book, I feel as vulnerable as if I had confessed a shameful secret; no longer can I employ the subterfuge of waiting until a second or third date to reveal where I attended college and what I studied.

The solution offered by happily married physicists such as Meg Urry is that female scientists should marry other scientists. One woman who followed this path wrote that while her friend the day-care provider saw the same moms day in and day out, she was afforded a vast array of male colleagues with whom she shared a common passion. Moreover, the men she meets tend to be starved for female company. But the high ratio of male to female scientists can set up an unsettling dynamic, as demonstrated by a letter from a male professor at a top engineering college who was appalled to learn that on his campus, where females make up only a quarter of the population, the men refer to something they call "Ratio-Induced Bitch Syndrome." Supposedly, women can act haughty and still find a boyfriend because they are in such high demand. In his twenty-one years on the faculty, this professor said, he has never seen evidence of this behavior. And yet, the slur is still out there, exactly the kind of gender-oriented peer pressure he sees as driving women out of STEM.

Several readers wrote to ask if women might feel fewer pressures if they attended all-girl high schools and universities. Given the testimony of female readers who graduated from single-sex institutions, this possibility is hard to ignore. A woman who matriculated at Wellesley College said what she and her friends in science and math learned was how free they felt to be whomever they wanted to be. She majored in chemistry and never had any reason to feel self-conscious about her success or to doubt her goals. At graduation, she and her fellow science majors all seemed to have a plan for themselves, whether this entailed a career in cancer drug research, chemical engineering, chemistry, applied math, or medical school.

When I remember the ostracism I experienced in junior high, I'm not sure I would have wanted to attend an all-girl school. Both sexes have a great deal to learn from each other, and boys who grow up without girls might see women as fetishistic objects of desire rather than classmates who understand circuits better than they do. And yet, the question seems worth studying, especially because countries where single-sex institutions are the norm report higher participation by female scientists. An astronomer wrote to say that when she began her career in the 1960s, many French physicists and astronomers were women. Three decades later, their percentages have dropped. A possible explanation? French high schools have become co-ed.

The observation that a woman's propensity to major in the hard sciences varies by culture seems borne out by readers who grew up in China, the former Soviet Union, India, and certain countries in South America, then immigrated to the United States and watched their daughters absorb the message that women can't excel in science or math. The complexity of cultural influences on women's participation in STEM fields was brought home to me when I was invited to attend an international conference of female physicists held in August 2014 in Waterloo, Ontario. Some of the women came from countries where their ability to excel in science wasn't doubted but they needed their husband's permission to attend any gathering at which men might be present. An older Iranian physicist, covered in a hijab, pointed out that girls in her homeland are raised not to dress seductively, so they don't feel the need to sacrifice their femininity to go into physics. And yet, even in Iran, where 47 percent of PhD students in physics are women, only 18 percent of the physics professors are female. A young Finnish physicist told me participating in any meaningful discussions at a Finnish conference is impossible for a woman because most such discussions take place in single-sex saunas.

Interestingly, several readers posited that European women find it easier to pursue careers in science because Europeans of both genders have healthier attitudes toward work than most Americans. Certainly, the unhealthy devotion required of graduate students in the United States seems designed to drive away as many women as possible. One reader said that despite graduating with a physics degree from a top university, she left graduate school because she resented being advised to spend fewer hours with her children for the sake of her research and to focus on her lab work over the holidays instead of visiting her family. Another woman said that even though her daughter earned a PhD in pure math from a highly ranked graduate program and published her dissertation in a prestigious journal, her advisor criticized her for spending too much time riding her bike and practicing yoga, saying she must not be devoted to math if she valued other aspects of her life. Rather than pursue a career in mathematical research, the daughter turned to math education, a field in which she felt more welcome.

Attitudes instilled in graduate school grow even more detrimental once scientists move on to run their own labs. Successful scientists argue and contradict each other constantly, one woman wrote. But girls are taught it's not nice to argue. Maybe, this woman said, we need to reconsider our propensity to teach girls not to interrupt, to always let everyone have a turn, to back down from any argument— behaviors parents are far quicker to discourage in a daughter than in a son.

Another woman demonstrated how the culture of science can drive away not only female researchers, but also men who care about their families. Her first serious summer job, in the 1980s, was in a chemistry lab where she was the only woman. The lab was decorated with photos of naked women. The men drank together after work and did not invite her to join them. If she had been determined to be a scientist no matter what, this wouldn't have dissuaded her from going on in chemistry. But she was uncertain, she said, and her sense of isolation was the deciding factor. Given that she was interested in several careers, why should she pursue a field that went out of its way to show her that she wasn't welcome?

Besides, she wrote, science doesn't even welcome men who care too much about their families. Her ex worked with a top scientist who often slept in his lab. When one of the grad students' wives went into labor, the student asked for a few days off. Clearly, the top scientist said, you aren't serious about your research.

While graduate school tends to be a stressful, isolating grind for everyone, women aren't imagining that the grind is more stressful and isolating for them. After my article appeared in the *Times*, Robert Leslie Fisher alerted me to his soon-to-be-published manuscript, "Invisible Student Scientists," in which he demonstrates that white male graduate students receive "more informal encouragement to start or continue research, more efforts to bring in speakers to talk about opportunities in the field, more financial assistance, more access to the latest equipment." In his earlier book, *Crippled at the Starting Gate*, Fisher cites data that female graduate students are less likely than men to be drawn into the inner circle of an advisor's favored students or to receive favorable teaching assignments.

As I expected, I heard from readers who grant that women are hindered by discrimination but wonder when Americans will start to care about the far more debilitating bias and lack of educational opportunities that plague minority students and those who come from poor or working-class families. One reader said that as much as she identifies with the women in my article, I failed to note the extra insanity that comes from being black and poor as well as female. As a student at the Bronx High School of Science in the 1990s, this young woman wanted to be an astrophysicist. She excelled in physics, chemistry, astronomy, and calculus, aced her AP classes, spent every summer attending engineering seminars, and came out in the top 5 percent of her class—despite living in the projects, having little money to buy materials, and needing to watch her siblings after school. She got into MIT but heard whispers from a few classmates that she had only made the cut because she was black. She found geochemistry to be less of an "old boys' club" than physics, but jobs in geochemistry offered low pay and demanded hazardous fieldwork in remote locales. Many women are unwilling to make such sacrifices, she wrote. And poor black women like her are *unable* to do so.

As much as I agree with such respondents—if I felt so out of place at Yale, I can barely imagine how isolated a poor black or Hispanic student might feel—my hope is that women and minorities might work together to improve the opportunities for all. The most gratifying response I received came from an African American engineer named Anthony Brown, whom I met when I showed up at our local radio station to do some interviews. He set me up in a studio, then slipped into his booth to oversee the technical aspects of the recording. When I finished, he came back in and hugged me. As a black man with a degree in computer science, he said, he constantly encountered the sort of bias I had been describing. Once, when he applied for a position, the person doing the hiring looked up from his résumé and said, *My, this is a very impressive list of accomplishments. Did you really do all this?* We traded battle stories, and I supplied him with statistics as to how many more calls for interviews a person with a white-sounding name receives compared to an applicant with the same qualifications but an identifiably African American name (the answer is 50 percent). I told him that in trying to convey the bias women and minorities face, I felt as if I were stating a fact as obvious as the sky is blue.

"Oh, sister," he said. "*You* may know the sky is blue. And *I* may know the sky is blue. But a whole lot of people out there do *not* know the sky is blue. You need to keep doing what you are doing. You are speaking for us all."

Still, the challenges faced by black, Hispanic, Asian, Native American—or gay—scientists aren't the same as those faced by straight white women. When my editor at the *Times* cut my article down to size, he eliminated any reference to women who weren't straight except for a parenthetical statement that lesbian scientists "reported differing reactions to the gender dynamic of the classroom and the lab, but voiced many of the same concerns as the straight women." I expected to hear from dozens of irate lesbians upbraiding me for glossing over the complexities of their experience; instead, I received a gentle remonstrance from a hydrologist named Vivian Underhill, who urged me not to neglect gay women's stories and thereby "miss out on a potentially illuminating discourse." If the *Times* had given her only one sentence in which to generalize queer women's experiences, Underhill wrote, "I'd say that these women hold at a minimum a double-minority status, and that this both magnifies the strength of their fears and also shifts them slightly."

In a column she used to write for a website called *Autostraddle*, Underhill drew on her career as a hydrologist to convey what it's like to be female and gay in engineering. In her first full-time job after college, she worked as a field hydrologist for the government. Even though she was out about her sexuality to her family and friends, she spent a lot of time feeling "weird and closeted and awkward" at work. "In our field, there's a lot of driving and hiking time, and therefore a lot of talking time; we get to know each other far better than many co-workers do. Where my co-workers (all men) could mention their girlfriends and fiancées offhandedly, I had to think and re-think about what I should say and how much I should allow." She only came out to her colleagues after a male coworker made an unwanted advance on a weekend rock-climbing trip.[1]

Underhill also alerted me to a study of lesbian, gay, and bisexual engineering students published in 2011 by Eric Cech, a sociologist at the University of California in San Diego, and Tom Waidzunas, a professor at Northwestern. After interviewing students at a major university, the authors concluded that even engineering students who don't hide their sexual preference may endure antigay sentiments, feel the need to downplay traits traditionally associated with their sexual identity, and keep quiet about the realities of their lives, often at the cost of "immense amounts of additional emotional and academic effort" and social isolation within an atmosphere the students already perceive as chilly.

According to Cech and Waidzunas, most engineers see their profession in binary terms: masculine versus feminine, technical versus social. Within this scheme, lesbians are perceived as more "masculine"

---

1. In a *New York Times* op-ed piece titled "Science's Sexual Assault Problem," published September 20, 2014, a professor of geobiology named A. Hope Jahren revealed that the sexual assault she had suffered as a graduate student doing fieldwork in Turkey caused her to change the focus of her research. Jahren cites a recent study by Kathryn B. H. Clancy and her coauthors, who surveyed 666 field-based scientists and found that 26 percent of the female scientists had been sexually assaulted while carrying out their fieldwork, as compared to 6 percent of the male scientists. "Most of these women encountered this abuse very early in their careers, as trainees. The travel inherent to scientific fieldwork increases vulnerability as one struggles to work within unfamiliar and unpredictable conditions." The perpetrators were predominantly senior members of the women's own research teams, and only 18 percent of those assaulted said they were aware of any way to report their harassment.

and therefore more competent as engineers than straight women, while gay men might be considered more "feminine"—and therefore less competent—than straight male engineers, although in both cases, being gay adds to the student's discomfort, an equation that becomes more complicated if an engineer is not only female and gay but also Asian, Latina, black, or Native American. (Asians, for instance, are seen as more technically proficient than Hispanics or blacks.)

What makes inequities in engineering so difficult to address is that anything related to feelings is deemed inappropriate. Such unspoken rules might be fine for straight, white male engineers, but the dismissal of social and political concerns keeps gay, lesbian, bisexual, or transgender engineers from talking about feeling isolated, discriminated against, or silenced. None of the students Cech and Waidzunas interviewed said they would feel comfortable discussing their concerns with a professor. One feared telling future employers she might need to find a job for her female partner. Another lesbian engineer characterized the engineers in her department as "straight dudes" who rarely consider the existence of people not like them and who think anyone who asks to be treated equally is trying to take some privilege away from them.

Surprisingly, the most combative response to my article came not from sexist male readers but from female scientists who strongly disagree that a job in academia allows so much flexibility that raising a family isn't hard. Several correspondents pointed out that finding a feminist husband isn't easy and finding an institution willing to hire a husband and wife can be a nightmare. Giving birth to a child precisely at the age when one is supposed to be carrying out the research required to earn a doctorate or tenure can be exhausting; quality day care is still hard to come by; and female scientists have a more difficult time cutting back their hours or bringing home work from the lab than women in business, medicine, or law. A woman with a PhD in molecular biology told me that she is applying for faculty jobs this year, even as she is expecting her first child. She worries about the inevitable obstacles to being a successful scientist and mother—for instance, the need for a lactation room near her lab, affordable day care, and a carefully structured plan so she can maintain her productivity even while on maternity leave. For every Meg Urry or Bonnie

Fleming who manages to combine motherhood with a high-powered career in physics, an uncounted number of women grow weary of struggling to succeed while carrying domestic burdens male scientists usually don't bear.

A female PhD who chose a career in industry rather than in academia took umbrage at Urry's claim that a job at a university allows more freedom. Although her PhD advisor had the flexibility to work from home, the advisor's workload was staggering. In contrast, her own schedule is more rigid, but when she leaves the office she has the freedom to turn work off. Her company provides on-site day care. And she earns more than she would as a professor, which lessens the stress of combining motherhood with a career.

An even greater contingent of readers wrote to complain that motherhood isn't the only reason a newly minted PhD would choose a career in industry—or medicine or law—over a career as a research scientist. To many, medicine looks far more appealing because doctors don't spend their lives begging for grants, exposing themselves to toxic chemicals, and sacrificing everything else in their lives to "publish or perish." The attrition rate is far lower for students in medical school than in graduate physics programs. Doctors can work part time, and their pay is far higher than academic scientists'.

These observations jive with the objections rendered by social scientists who question whether the shortage of STEM workers in the United States is real and, if so, why such a shortage hasn't resulted in higher wages. A Rutgers professor named Hal Salzman argues[2] that the alarm about the supposed lack of trained engineers in the United States has been generated by companies trying to justify shipping their operations overseas or pressuring Congress to ease restrictions on visas for students and guest workers. "If there were a talent shortage," Salzman writes, "where are the market indicators (namely wage increases) that signal students there is an opportunity to pursue a career in this industry that is better than the alternatives?"

While I admit a career in physics or mathematics might be a lousy choice for a young woman who could earn more money and exert

---

2. See "What Shortages? The Real Evidence about the STEM Workforce," published in the summer 2013 edition of *Issues in Science and Technology.*

more control over her life if she went to medical school or took a job in industry, anyone who loves physics or math ought to be able to pursue her passion without enduring so many slights she decides to become a doctor or gives up a promising career to stay home and raise her kids. Jobs in STEM fields offer better salaries and more rewards than most traditionally female jobs. The scientists of today are developing the next generation of computers upon which all of us will depend. They are designing the cars that will drive us from here to there and the robots that will operate on our bodies, mind our children, and nurse our elderly parents. Scientists studying global warming or deciphering the true nature of dark matter are laying the groundwork for policies that will shape our future and the philosophical vision that will determine our collective psyche. Shouldn't we include a fair proportion of women, gay people, and scientists of color among those exerting such an influence on how we live and our understanding of what it means to be human?

That researchers question the shortage of workers in STEM fields doesn't bother me. What shocks me is how many scientists offered heated yet utterly illogical responses to my article. One writer, "Markus from New York," commented on the *Times* website that Jo Handelsman's study, in which researchers of both genders showed a significantly greater tendency to hire and mentor an applicant named John—and pay him more money—than an identical competitor for the job named Jennifer, did not demonstrate a bias against female scientists. According to Markus, an employer might be making such a judgment based on his *experience*, which "objectively showed" that women, on average, are less able than men.

Such arguments sound remarkably logical, and yet, the practice Markus defends is the very definition of bias. No study has proven women to be less talented, less effective, less productive, or less committed scientists than men. Even if someone were to believe his "experience" justified a decision to hire a man, he would be making that decision based on nothing but his feelings, or a few anecdotal examples, which is what prejudice is all about. (Imagine the same employer justifying his decision to hire a white scientist based on his "experience" that black people are rarely as intelligent, hard working, productive, or committed as white people.) Also, if the scientists who hired John were basing their decisions on experience, wouldn't the study have recorded a difference between the hiring practices of younger and older scientists? Wouldn't younger scientists (especially females) be more innocent and idealistic and therefore more likely to view women and men as equals than scientists basing their decisions on decades of "experience" to the contrary? And yet, the data show no such difference.

When I asked Handelsman about such criticisms, she confirmed that prejudice is indeed defined as making judgments about an individual based on group characteristics, whether these characteristics be imagined or real. "The problem with such prejudices is that whether the generality is true or not, the individual case may be different," she wrote. "The employer may hire a man and be surprised when he announces he's going on parental leave three months after starting work. Or he may hire the woman and find out she has no intention of having children and is the best employee he's ever had. . . . The point is that we miss characteristics about people when we judge them by group characteristics." Handelsman challenged anyone to produce data that women "actually are less productive, despite taking time for children. My most productive graduate student EVER had one child, a miscarriage, and another pregnancy while in graduate school and finished in less time than any other graduate student I've had in 28 years. So I think it's a convenience to reject women because of child bearing when we don't measure how much time men take off for illness, how much time they spend at work talking sports, and their willingness to respond to criticism."

The question of female productivity raises the question of how such productivity might even be measured. Virginia Valian, in her comprehensive and highly readable book *Why So Slow? The Advancement of Women,*[3] cites evidence that male academics, on average, publish a greater quantity of papers, but females publish papers of higher quality. Moreover, the higher rate of publication for males may also be due to bias. The review process at most journals is far from blind.

---

3. *Why So Few, Why So Slow* . . . the titles of books about women in science read like a prayer for patience.

Still, when every other explanation fails, supporters of innate differences resort to the circular argument that the fact that so many more men than women end up as physicists and mathematicians proves women prefer fields other than physics and math. One correspondent pointed out that my success as a writer proves I chose not to become a physicist because my real passion lay in writing, while others claimed women choose not to become physicists because they care more about nurturing living things, an argument beautifully countered by the scientist who wrote that she loves spending her days working with inanimate objects because she is expected to spend nearly every other hour of her day taking care of living creatures.

According to Urry, the key to reforming the system is convincing scientists that broadening the pool of female students and faculty members and making the culture more livable for both genders doesn't lower standards. Rather, you are dumbing down science if you rely only on men. If society needs a certain number of scientists, and you can only look for those scientists among the white males of the population, you are going to need to dig much farther toward the bottom of the barrel than if you could also search among women and scientists of color.

Robert Leslie Fisher argues that around-the-clock dedication might have been required for scientists racing to develop the atomic bomb, but a more humane and balanced approach would make for better science in peacetime. Even in cutting-edge fields in which researchers can sense competitors breathing down their necks, a more cooperative ethos could produce faster, more innovative results. Consider the military, Fisher says. You want marines with the physical strength to pull a comrade into a foxhole. But you also want soldiers who can invent missile defense systems, break the enemy's codes, manage a deployment, and win civilian hearts and minds.

Rather than trying to weed out young physicists prematurely, why not help students who didn't receive a sterling preparation in high school make up for what they missed? At some point, some students will discover they aren't talented enough to succeed at physics, or they aren't passionate enough about a career in science to forego the higher salaries they might earn in medicine, or they would prefer to write novels or join the Peace Corps. But why not help as many undergraduates as possible remain in the major, then encourage the most talented to go on to graduate school? Would productivity really go down if graduate students weren't pressured to remain in the lab until they became too bleary-eyed and muzzy-headed to think? The standards in medicine didn't suffer when interns were allowed to get some sleep.

Those in charge need to make it easier for researchers of both genders to raise a family by slowing the tenure clock or reducing the teaching load of young parents, providing on-campus day-care facilities, and taking into account a parent's schedule when assigning teaching duties and committee work. Faculty and administrators need to be aware of the effects of implicit biases on hiring and promotion. At the University of Michigan, anyone who serves on such committees is required to sit through workshops on the benefits of broadening the pool of candidates, judging all applications on the same basis, making sure to include more than one woman or minority in the cadre of applicants who are brought to campus, and not asking female candidates about their spouses or their plans to have children.

Timothy McKay, an astrophysics professor at Michigan, sees great value in workshops designed to counter unconscious bias in hiring. With other members of the physics department, he is implementing programs to counter the mindset that prevents female students from believing they can do science or math. "It is quite common for an office hour visit from a student to begin with a disclaimer like 'I'm not a physics person,'" McKay wrote me. "If you're 'not a physics person,' there's no particular reason to try hard (it won't work anyway) and no great inner cost for failure (it's just not your thing)." At Michigan, the physics department is helping female students and minorities overcome such psychological barriers, most notably through a computer program tailored to provide personalized feedback, encouragement, and advice to each of the 1,800 students who take physics each term. Not long ago, McKay won a $2 million grant to change the culture of teaching in introductory STEM courses, improve the climate in these departments, and reduce the gap in performance between male and female students.

. . .

As feminists learned in the 1960s, if women compare their experiences, such conversations might generate scientific theories that can be tested by studies, as well as the realization that a woman's frustration might not be the result of any lack of talent or dedication. I don't want to dismiss a male scientist's anecdotes about the women in his lab being treated equally to the men by pointing to studies that document bias against women, only to privilege a female scientist's anecdotes of discrimination. But examples of discrimination can prove discrimination exists in ways that examples of women or people of color being treated fairly cannot prove discrimination does *not* exist. Some aspects of human behavior can be understood only by means of narrative—that is, the detailed communication of the particularities of an experience that illuminate the psychology and actions of a group.

Only by comparing experiences can patterns be discovered and changes in the society initiated and carried out. After reading a dozen e-mails from readers who identified with the graduate student and postdocs at Yale who described themselves as "the women who don't give a crap," I considered starting a movement by that name. As one male reader wrote, gender bias and jellyfish stings didn't prevent Diana Nyad from fulfilling her dream to swim from Cuba to Florida. And yet, if the only female scientists who attain positions of power are superwomen like Diana Nyad, those few who succeed might not see anyone less heroic as deserving of help. A number of female scientists complained to me that older women are sometimes the worst perpetrators of bias and abuse against their younger female colleagues. Older female scientists might be harsh with their younger colleagues because they are trying to toughen them up to survive the hazing the older women received. But just because one generation of women is abused, they needn't pass on that harassment.

So, yes, women might need to change. They might need to become more confident, less needing of encouragement, more eager to support each other. But if female scientists are weighted with the responsibility of mentoring younger female scientists, they will have even less time to carry on their research. It's the larger society that needs to change. No American of either gender will want to become a scientist if studying science or math makes a middle schooler so nerdy he or she becomes undatable, or if science and math are taught in such a way as to seem boring or irrelevant. Focusing on facts and tests is not the best way to convey the beauties of a subject or the reasons anyone would want to study it. The lowly status teaching is accorded in our society, combined with the unreasonable demands on most teachers, makes it difficult for anyone to impart high-level skills to our children while instilling in them a love for whatever is being taught. Changing such deeply ingrained cultural patterns might be difficult, but the barriers are not insurmountable. Optimally, the new Common Core curriculum will help to achieve this goal, but only if teachers are trained to understand science and math at the deepest levels.

Luckily, programs are being developed to attract young women and minorities into STEM. According to one reader, in 2001, thirty women at Texas Instruments founded the High-Tech High Heels Fund. After the group trained guidance counselors and teachers in Dallas to encourage female students to enroll in more challenging classes, the number of AP Physics tests taken by Hispanic, African American, and female students more than doubled. The fund also sent hundreds of girls—most of them minorities, many of them poor—to a two-week AP Physics camp, with the result that the girls achieved a much higher success rate on the AP Physics exam than they otherwise might have done.

The female owner of a high-tech start-up wrote to describe her frustration at trying to convince women they can learn to write computer code . . . and her joy at witnessing their eventual success as programmers. Even small changes can make a big difference: a thoughtful male professor suggested changing the name of the major from Computer Engineering to Computer Arts, a switch he and I are convinced would double the number of women overnight.

Dropbox, which employs mostly white and Asian men, is working to attract more women and minorities. According to an article in *USA Today* on November 6, 2014, "Job interview questions about a zombie apocalypse, and conference rooms named the 'Bromance Chamber' and 'The Breakup Room' didn't help the company's reputation for being a boys' club." But Dropbox trained its recruiters to detect their biases, and in less than ten months, the company doubled the percentage of women in engineering and tripled the percentage of women in design. An African American account manager named

Justin Bethune led the company's first tour of historically black colleges and universities and reported interest levels to be "off the charts from both students and administrators. We had students lining up to ask us what it was like to work at Dropbox and tell us about how Dropbox is a part of their daily lives, and administrators kept wanting to make sure we'd come back." In January 2015, Intel announced a similar campaign, setting aside $300 million to attract and retain women and minorities and to support engineering scholarships, especially at historically black schools. The money also would be used to encourage women to enter the gaming business.

For all that, the majority of the responses I received lead me to think that change will come simply because people are growing more aware of the cultural and psychological biases that until now prevented women from achieving all they might achieve. Margaret Magnus, a software engineer and former owner of an IT business, wrote to say she felt a "wonderful lightness" on reading my article. "It's easier to endure something, and also to defend yourself against it, to the degree that you are free of doubt that it is in fact happening, and that it's not okay."[4]

If you still doubt the power of self-awareness, witness this final message, from a mother whose daughter earned a grade of 100 percent in her first-year calculus course but was struggling with the honors math class that followed. The mom, an avowed feminist, worried the course was too hard and berated her engineer husband for projecting his ambitions onto their daughter. She urged their daughter to drop the class.

---

4. One caveat. An essay by Adam Grant and Sheryl Sandberg, published in the *New York Times* on December 6, 2014, warns that if you alert people to the tendency to stereotype women and then ask them to make a hiring decision, the bias against women *rises*, possibly because the managers absorb the message that bias must be socially acceptable or justified. Rather, the message needs to be that bias is common but most people want to conquer their prejudices and doing so is beneficial. With such a warning, "managers were 28 percent more interested in working with the female candidate who negotiated assertively and judged her as 25 percent more likable." Similar findings show if women are told few females apply for certain positions, the numbers don't change. Only if management adds "We want this to change!" do the numbers shoot up.

Her husband, however, stuck to his guns. "He said failing grades were just fine and no indication of anything. He told her to work harder and longer and take advantage of the resources available (which are considerable, in the internet age). He did some hand-holding. My brother—with a degree in math and computer science—said the same, and just shook his head at me.

"When I read your story, I recognized myself and was appalled. I immediately sent it to my daughter and let her write the midterm without further comment."

As a result, the woman said, her daughter ended up as one of the top ten students in her class, with renewed confidence and determination. She intends to stick with honors math. At the very least, her mother said, "I think she now knows she belongs there if she chooses to stay."

ACKNOWLEDGMENTS

Not only did Meg Urry give generously of her time in sharing her insights and experiences as a female physicist, she also read an early draft of this manuscript. This book (and my life) would be far poorer without her contributions.

I am deeply appreciative of the wisdom and support I received from the following people: Bonnie Fleming, Jo Handelsman, Corinne Moss-Racusin, Judith Krauss, Yevgeniya Zastavker, Leslie Klein, Erika Karplus, Jedidah Isler, Eilat Glikman, Ivy Wong, Elizabeth Jerison, Shana Elbaum, Tatenda Shoko, Adele Plunkett, Judy Dixon, Alejandro Uribe, Cinda-Sue Davis, Jamie Saville, Timothy McKay, Abigail Stewart, Terrence McDonald, Anne Curzan, Anita Norich, Peggy Burns, Vasugi Ganeshananthan, Jeri Finnegan, Jack Strassman, Michael Hazelnis, Lucinda Nolan, Marie Vitulli, Kate Sher, Janet Malachowsky, and Rochelle Sharpe.

Although they did not wish to be identified by their full and/or real names, a great many of my friends and former classmates, as well as students at Yale, Liberty Central High School, the University of Michigan, and other institutions, helped to make this a more complex, honest, and nuanced book than otherwise would have been the case.

I would never have pursued my love of science and math as deeply as I did if not for my teachers: Ed Wolff, James Gallagher, Roger Howe, Michael Zeller, Peter Parker, and Peter Nemethy; they also were extremely kind and openhearted in allowing me to interview them. I would not have survived four years at Yale without the friendship of Laurel Omert. I hope my gratitude toward Barry Talkington and John Hersey is adequately expressed in the body of this manuscript; I would never have become a writer without them. Nor would I have survived the ordeal of writing this book if not for the emotional support and editing suggestions provided by Maria Massie, Therese Stanton, Doug Trevor, Susan Hildebrandt, Marian Krzyzowski, Laura Kasischke, Suzanne Berne, Maxine Rodburg, Michele McDonald, Marcie Hershman, Peter Ho Davies, and my son, Noah Glaser.

To my editor at Beacon Press, Amy Caldwell: thank you for understanding what I was trying to do and helping me to integrate my scientist and writing selves; added thanks to everyone at Beacon who worked so hard and with so much enthusiasm to bring my book into the world.

Like so many others, I can never repay my debt to Gloria Steinem and all the other brave feminist thinkers, writers, scientists, and activists who came before me.

Nothing in this book is invented; my experiences are recollected to the best of my ability, verified whenever possible by interviews with other participants and consultation with my diaries, journals, and other writings. Any errors are unintentional and the result of the normal limitations of human memory. All quotations are as accurate as I can make them—again, based on memory, notes, and other written records. I have changed names and identifying details only as indicated, whether to protect interviewees who asked me to shield their identities, or to spare professors and other people from my past who did nothing to warrant being singled out for censure. In no case have I created a composite figure.

PREFACE: BRIGHT COLLEGE YEARS

For more details as to the content of Summers's remarks, see Patrick D. Healy and Sara Rimer, "Furor Lingers as Harvard Chief Gives Details of Talk on Women," *New York Times*, February 18, 2005.

For statistics on the number of women and minorities receiving degrees in physics and computer science, see online sources such as the American Physical Society, http://www.aps.org/programs/women/resources /statistics.cfm, and the National Center for Women and Information Technology, http://www.ncwit.org/sites/default/files/legacy/pdf /ByTheNumbers09.pdf.

For the study documenting the ways in which responses to students' requests to meet with a professor differ by gender and race, see Katherine L. Milkman, Modupe Akinola, and Dolly Chugh, "What Happens Before? A Field Experiment Exploring How Pay and Representation

Differentially Shape Bias on the Pathway into Organizations" (working paper, Social Science Research Network, April 23, 2014), http://papers .ssrn.com/sol3/papers.cfm?abstract_id=2063742.

CHAPTER TWO: SCIENCE FAIR
For the audio version of the exhibit at the Travelers Insurance Pavilion at the 1964 World's Fair, see http://www.nywf64.com/travelers05.shtml.
The ad for the small space ship comes from *The Wonderful Flight to the Mushroom Planet* (New York: Little, Brown, 1954); the quotes about David and Chuck and their skill at building such a ship come from *Stowaway to the Mushroom Planet* (New York: Little, Brown, 1956). The author of the series is Eleanor Cameron.

CHAPTER THREE: SCIENCE UNFAIR
Ted Mooney's website about electroplating can be found at Finishing.com, http://www.finishing.com/faqs/howworks.html.

CHAPTER FIVE: FRESHMAN DISORIENTATION
For a more detailed description of the Title IX protest by the Yale women's crew team, see the *Yale Daily News*, http://yaledailynews.com/magazine /2002/11/14/title-ix/.
My copy of *Physics* by David Halliday and Robert Resnick was published in 1960 by Wiley. The three-volume *Feynman Lectures on Physics* was originally published in 1970 by Addison Wesley Longman. A new "Millennium Edition" is currently for sale, although the text also seems to be available for free online. For an in-depth look at Feynman's life and mind, see James Gleick, *Genius: The Life and Science of Richard Feynman* (New York: Vintage Books, 1993). *Feynman's Tips on Physics* was edited by Michael Gottlieb and Ralph Leighton and published in 2005 by Addison Wesley; most of the examples I cite come from pp. 51–53.

CHAPTER SEVEN: ELECTRICITY AND MAGNETISM
For videos of the Chladni plate experiment done correctly, see YouTube: http://www.youtube.com/watch?v=n-tYVjngvyo.

CHAPTER EIGHT: THE PHILOSOPHY OF EXISTENCE
My article on Jacob Bronowski appeared in the April 1976 issue of the *Yale Scientific*, under the title "Jacob Bronowski: Modern Renaissance Man."
Beverly Berger went on to an illustrious career as a researcher in the field of theoretical gravitational physics. She taught for many years at Oakland University in Michigan, where she eventually became chair of the Physics Department, and later served as the program director for gravitational physics at the National Science Foundation. In an e-mail dated

March 17, 2014, she told me, "My husband and I will celebrate our forty-fourth wedding anniversary in August. We have no children."

CHAPTER NINE: X-10, Y-12, K-25

My column on my experience at Oak Ridge appeared in the November 1977 issue of the *Yale Scientific*, under the title "Inside the Oak Ridge Fence."

CHAPTER TEN: LIFE ON OTHER PLANETS

Gravitation was written by Charles W. Misner, Kip S. Thorne, and John Archibald Wheeler; my copy was published in 1970 by W. H. Freeman.

"Women Plus Math at Yale: Sometimes It Doesn't Add Up" was published in the *Yale Daily News* on December 1, 1976. "Confessions of a Faltering Physics Student" appeared in the October 1977 issue of the *Yale Scientific*. "A Passion for Propagators," my essay about my senior project on the propagation of waves in n-dimensional space, was published in the *Yale Scientific* in February 1978. "Lucy, Gracie, Marie, and Me" was published in the *Yale Scientific* in March 1978.

The four-colour theorem was proven by Kenneth Appel and Wolfgang Haken in 1976.

CHAPTER ELEVEN: THE TWO-BODY PROBLEM

My return to New Haven and the interviews recounted in this chapter took place in the fall of 2010.

"That's Not My Job: Zeller's Laws of Attraction" was written by Amanda Lewis and appeared in the *Yale Daily News* on February 12, 2007.

For a discussion of the attempt to revamp introductory science and math classes, see Richard Pérez-Peña, "Colleges Reinvent Classes to Keep More Students in Science," *New York Times*, December 27, 2014.

In addition to Meg Urry's February 6, 2005, op-ed piece in the *Washington Post* ("Diminished by Discrimination We Scarcely See"), I have also drawn on her essay "Why So Few? How to Increase the Number of Women in Science," which appeared in the *RHIC News*, the newsletter of the Brookhaven National Laboratory's Relativistic Heavy Ion Collider, on June 3, 2008.

CHAPTER TWELVE: STATICS AND DYNAMICS

My interviews with Jeri Finnegan, Jack Strassman, Cindy Nolan, Mike Hazelnis, the girls in Cindy Nolan's physics class, and Arthur Olson took place in Liberty, New York, in the fall of 2009.

When I went back to the library at Sullivan County Community College, I found the same books on the shelves: A. S. Eddington's *The Nature of the Physical World* (Ann Arbor: University of Michigan Press, 1958);

and Erwin Schrödinger's *Science, Theory and Man* (1935; repr., New York: Dover Publications, 1957).

CHAPTER THIRTEEN: INTEGRATION AND DIFFERENTIATION
The book I used to teach myself calculus was *Calculus and Analytic Geometry* by George B. Thomas Jr., published in 1951 by Addison-Wesley, now in its twelfth edition. The Dummies series is published by Wiley; *Algebra I for Dummies* (2001) and *Algebra II for Dummies* (2006) were written by Mary Jane Sterling, as was *Trigonometry for Dummies* (2005; the joke about the squaws and the hippopotamus hides is on p. 87). Mark Ryan is the author of *Geometry for Dummies* (2008).

Danica McKellar's book *Math Doesn't Suck* was published in 2008 by Plume, which also published the sequels *Kiss My Math* (2009), *Hot X* (2011), and *Girls Get Curves* (2013).

I am grateful to Ed Wolff and James Gallagher not only for teaching me mathematics when I was in high school—most especially Jim for loaning me his calculus book and trying to help me with my studies—but also for allowing me to interview them in the fall of 2009, especially given Jim's illness.

CHAPTER FOURTEEN: THE WOMEN WHO DON'T GIVE A CRAP
I spoke with Jamie Saville and Cinda-Sue Davis in October 2010; my telephone interview with Carrie took place in November 2009.

The results of the study by the American Mathematical Society can be found in Titu Andreescu et al., "Cross-Cultural Analysis of Students with Exceptional Talent in Mathematical Problem Solving," *Notices of the American Mathematical Society* 55, no. 10 (November 2008): 1248–60, http://www.ams.org/notices/200810/200810-full-issue.pdf. A discussion of the study was also published in Sara Rimer, "Math Skills Suffer in U.S., Study Finds," *New York Times*, October 10, 2008.

A description of the 1980 study of mathematically precocious American middle schoolers by Julian C. Benbow and his colleagues is to be found in the AMS study already cited; as well as in C. P. Benbow and J. C. Stanley, "Sex Differences in Mathematical Ability: Fact or Artifact?," *Science* 210, no. 4475 (1980): 1262–64, and at http://www.vanderbilt.edu/Peabody/SMPY/ScienceFactOrArtifact.pdf.

The AAUW's 2010 report *Why So Few?*, written by Christine Hill, Christianne Corbett, and Andresse St. Rose, can be downloaded at http://www.aauw.org/learn/research/whysofew.cfm.

The claim that girls and boys perform equally well in science and math in elementary school is substantiated by research cited in *Why So Few?*, as well as many other sources. The data for the percentages of girls among all high school students taking physics, math, and computer science can be found at the American Institute of Physics website, http://www.aip .org/statistics/trends/reports/hsfemales.pdf.

See Londa Schiebinger, *Has Feminism Changed Science?* (Cambridge, MA: Harvard University Press, 1999), 39, 55, 58.

For details of the Michigan study of male and female performance on math exams, see Steven J. Spencer, Claude M. Steele, and Diane M. Quinn, "Stereotype Threat and Women's Math Performance," *Journal of Experimental Social Psychology* 35 (1999): 4–28. Other such studies are discussed on p. 39 of *Why So Few?*; the discussion of Shelley Correll's research comes from p. 44 of *Why So Few?*

The study of attitudes toward math and science among Yale undergraduates was carried out by Brian A. Nosek, Mahzarin R. Banaji, and Anthony G. Greenwald and is described in "Math = Male, Me = Female, Therefore Math ≠ Me," *Journal of Personality and Social Psychology* 83, no. 1 (2002): 44–59.

My discussion of David Anderegg's work comes from his book *Nerds* (2007; repr., New York: Tarcher/Penguin, 2011), as well as e-mail communications with the author.

The quotation from Melanie Wood comes from Rimer, "Math Skills Suffer in U.S.," as does the statement from the Romanian math champion.

For a discussion of the rate of production of female physicists in other countries as compared to America, see p. 39 of Schiebinger's *Has Feminism Changed Science?*, as well as http://www.aip.org/statistics/trends /reports/women05.pdf. See also Robert Leslie Fisher, *Crippled at the Starting Gate* (Lanham, MD: University Press of America, 2010), 12. The information about Richard Feynman's IQ comes from p. 30 of Gleick's biography.

All discussions of the work of Catherine Riegle-Crumb come from private communications with the author. For a list of citations, see her CV at http://www.utexas.edu/cola/centers/prc/directory/faculty/cr653.

The statistic about the SAT math scores of white males in engineering, computer science, math, and the physical sciences in America can be found on p. 21 of *Why So Few?*

My interviews with Travis and Mindy (not their real names) occurred in the fall of 2010, as did my interview with Bonnie Fleming and my dinner with the five female scientists in Harvard Square.

For details of the MIT study and the quotes in this section, see the *MIT Faculty Newsletter* online, http://web.mit.edu/fnl/women/women.html. The quote from Nancy Hopkins can be found at the *Nature* website, http://www.nature.com/scitable/forums/women-in-science/nancy-hopkins -on-mit-for-women-in-18883712.

Judith Kleinfeld's rebuttal of the MIT study can be found at *Massachusetts News*, http://www.massnews.com/past_issues/2000/4_April/aprmit.htm.

A copy of the 2012 survey of resources available to male and female physicists, conducted by the American Institute of Physics, is available online at http://www.aip.org/statistics/reports/global-survey-physicists.

The study by Jo Handelsman, Corinne A. Moss-Racusin, and their colleagues (John F. Dovidio, Victoria L. Brescoll, and Mark J. Graham) was published in "Science Faculty's Subtle Gender Biases Favor Male Students," *Proceedings of the National Academy of the Sciences* 109, no. 41 (October 9, 2012): 16474–79, http://www.pnas.org/content /109/41/16474. I carried out my interviews with Handelsman and Moss-Racusin in person and via e-mail and telephone from the fall of 2012 through the fall of 2013. The chart showing the results of the study was provided by Moss-Racusin.

For a discussion of the study by Lubinski and Benbow, see the Vanderbilt University website, http://www.vanderbilt.edu/peabody/smpy /DoingPsychScience2006.pdf.

In the op-ed essay "Academic Science Isn't Sexist," from the November 2, 2014, edition of the *New York Times*, Wendy M. Williams and Stephen J. Ceci argue that all studies indicating bias in the sciences are old or anecdotal and that their own study "reveals that the experiences of young and midcareer women in math-intensive fields are, for the most part, similar to those of their male counterparts." The real problem, the authors assert, is that women "choose" or "prefer" to work with people or animals; if only they could be lured into STEM fields, they would find gender equity and job satisfaction. Not only do Williams and Ceci ignore Handelsman's 2012 study proving gender bias, the validity of their own study has been called into question, most notably by Rebecca Schuman, the education columnist for *Slate*, and science bloggers such as Jonathan Eisen, P. Z. Myers, and Emily Willingham, who point out the many inequalities that Williams and Ceci's own study seems to document.

The 2012 report of the President's Council of Advisors on Science and Technology can be found at the White House website, http://www.whitehouse .gov/sites/default/files/microsites/ostp/pcast-engage-to-excel-final_feb.pdf.

My visit to the Franklin W. Olin College of Engineering and my interview with Yevgeniya Zastavker and her students took place in September 2010. For a discussion of the research involving women's performance on tests of spatial aptitude, see pp. 52–56 of *Why So Few?*, most especially the work of Sheryl Sorby, a professor of engineering at Michigan Technological University who has studied the importance of spatial-skills training to the retention of women in engineering.

My meeting with Leslie Klein in New York took place in September 2010, as did the picnic I attended for the Yale Physics and Astronomy Departments; I conducted my telephone interview with Erika Karplus in August of that same year.

CHAPTER FIFTEEN: PARALLEL UNIVERSES

The quote by Andrea Dupree can be found on p. 169 of Schiebinger's *Has Feminism Changed Science?*

The figure about 4 percent of the American workforce comes from p. 2 of *Why So Few?*

The quotation from Larry Summers comes from a personal communication via e-mail dated July 14, 2011. I am deeply grateful for his willingness to read an early draft of this manuscript and for his support for my project.

Both the *Yale Daily News* and the *New York Times* ran stories on April 13, 2011, about Michele Dufault's tragic accident. The rest of my knowledge about Michele and her death come from private e-mails from Meg Urry sent later that month.

EPILOGUE: THE SKY IS BLUE

I highly recommend Virginia Valian's *Why So Slow? The Advancement of Women* (Cambridge, MA: MIT Press, 1998), as well as Vivian Gornick's *Women in Science: Then and Now* (New York: Feminist Press, 2009), two seminal works I somehow failed to discover until after they had been pointed out to me by readers of my article in the *New York Times*.